SAKEPEDIA

A NON-TRADITIONAL GUIDE
TO JAPAN'S TRADITIONAL BEVERAGE

By

JEFF CIOLETTI

TURNER PUBLISHING

Turner Publishing Company
424 Church Street • Suite 2240 • Nashville, Tennessee 37219
445 Park Avenue • 9th Floor • New York, New York 10022
www.turnerpublishing.com

Sakepedia

Cover design: Maddie Cothren
Book design: Nathalie Ouederni

Library of Congress Cataloging-in-Publication Data

Cioletti, Jeff.
Sakepedia / A Non-Traditional Guide to Japan's Traditional Beverage by Jeff Cioletti.

 pages cm

1. Alcoholic beverages--History.
2. Drinking of alcoholic beverages.
I. Title.
TP520.C56 2015
641.2'1--dc23
2015009483

[[9781630267582]]

Printed in the United States of America
15 16 17 18 19 20 10 9 8 7 6 5 4 3 2 1

CONTENTS

INTRODUCTION

Many adjectives have been used to describe sake over the years, but the one I admittedly find most annoying is *exotic*. There's nothing wrong with the word, per se, nor is there anything wrong with those who apply the term to a beverage. I just find "exotic" to be a sort of shorthand for, "What is this strange, mysterious liquid and why am I so intimidated by it?"

And that's been more of a marketing problem than a consumer problem. When you walk into a shop with a sizeable stock of the fermented rice beverage, it's often difficult to distinguish one from another. The bottles are practically identical, and the labels are covered in Japanese kanji, which most of us can't decipher. Sure, there's usually some English, but you really have to look for it. The issue, it seems, is a lack of branding.

I spoke about this marketing void with Steve Vuylsteke, CEO of SakéOne, one of the few US-based breweries producing premium-grade sake as well as an importer of select brands from Japan. Vuylsteke observed that there's a bit of disconnect between American and traditional Japanese concepts of branding.

"I had one Japanese brewer tell me that their idea of marketing is a bit more antiquated, maybe a bit more production-driven," Vuylsteke says. "In other words, make the sake and hope people like it, versus figure out what the consumer needs, make a sake to fit that need, and make a label to reflect that. We're already dealing with a country that has very little knowledge of the category, so you can't just shove a bottle that's all kanji in front of people and expect them to know what it is. There's still some resistance to that, but I think it's starting to change."

Retailers aren't completely innocent in this either. A mega alcohol superstore might devote ten horizontal and five vertical feet to the drink (which is impressive for any venue that doesn't specialize in it), but sake is still, for the most part, ghettoized. In fact, next time you're in one of those mondo booze retailers, head to the wine section and try this experiment. Find the most esoteric varietal and ask the salesperson for a recommendation. Chances are, that person will be more than eager to go into great detail on flavor and aroma notes, terroir, vintage,

1

food pairing suggestions, and the head winemaker's alma mater (okay, maybe not that last part).

It's likely that the shop also will have an extensive collection of craft beers and someone on staff who can wax poetic about New England IPAs, barrel-aged barley wines, and what's the hottest new hop variety from New Zealand (that wouldn't have been the case a decade ago, but, let's face it, beer has finally arrived).

Now find that more modest, but still generous, slice of real estate that holds the sake and ask for assistance. At worst, you'll get a blank stare, and at best, you'll get a response like, "Uh, it's made out of rice."

But I don't want to throw retailers under the bus. Understandably, they're focusing their educational resources on the beverage categories that pay the bills. And how many consumers are really engaging them with questions about sake on any given day?

Unfortunately, that's a product of the vicious cycle that is intimidation. Sake is so unfamiliar to many consumers that they're likely to steer clear of it. The non-specialty stores will continue to stock the beverage because there is a small but consistent group of aficionados and sake-curious shoppers who make it financially viable for those outlets to carry it. But there's little hope of expanding beyond that base because the salespeople aren't going out of their way to tout the category—mainly because customers are too intimidated to develop curiosity about this odd, mystical rice elixir.

Part of my goal with this book is to stoke curiosity among those who have had limited exposure to sake, as well as to enrich the experience for those who've dipped their toe in or fully immersed themselves in the beverage (not literally, of course—then again, who am I to judge?).

Some level of detail regarding the production process of sake is necessary to enhance one's appreciation of the drink, but not on a total geek's level. Trust me, I keep it pretty painless.

One thing I've never been a fan of in the drinking world is pretension. Sadly, it's all too present in many beverage categories, and it's a pitfall most of us try to avoid when we start to flirt with connoisseurship of any kind (I'm looking at you, wine!). Yes, there are tasting notes in this book. I review hundreds of different sakes from brewers all over Japan as well as a few from North America and a couple other parts of the world. Actually, *describe* is a better term, because I'm not passing any sort of judgment on what I'm tasting. Just assume that if it's in this book, it's good.

And besides, "good" is so subjective—as is the act of tasting, for that matter—and who the hell am I to tell you what you should or shouldn't like?

Which brings me to my next point. I have a bit of a confession to make: in the early days of my better-drinking odyssey, I may have, on more than one occasion, wandered a bit into snob territory. It was more pronounced during the first couple of years of my relationship with craft beer. But it did rear its head a little bit with sake—mainly as it pertains to so-called "sake bombs." You've seen them. They're not unlike Boilermakers, except instead of dropping a shot of whiskey in a glass of beer, you're doing the same with a similar volume of sake. I'm not an advocate of such a practice. Usually when someone asks for my thoughts on that dubious ritual, my advice is pretty simple: "Don't." But I'm not nearly as militant about it as I once was (if you read my first book, *The Year of Drinking Adventurously*, you'll know what I'm talking about). While I still choose not to partake, I have absolutely no judgment for those who do. If it means one more person is going to discover sake—regardless of the fact that it's overwhelmed by a totally different beverage—it would be silly for me to try to stand in the way of that experience. Just don't expect to find any reviews on which bars make the best bombs or suggestions on what dishes pair best with the beer-sake cocktail.

As for the descriptions that will appear in this book, I try to make them as tangible as possible. Wherever possible, I compare aroma and flavor notes to foods you're likely to have smelled or tasted rather than something otherworldly. You're going to see a lot of "apple," "pear," "cooked rice," "cherry," and "nuts" throughout these pages. I want the vocabulary to be meaningful and evocative and not something that's going to alienate the average reader. I once took a tasting class where one of the other students kept using the term *river rock* to describe what he was experiencing. Eventually I said to him in a fairly good-natured way, "You and your f*!@ing river rocks!" He explained that he came from the wine sommelier world and those are the types of words you're expected to use among your peers.

If there's anyone out there who, while on a hike, white-water rafting trip, or other let's-commune-with-nature activity, has ever picked up a stone from a stream and licked it, please raise your hand. I'll wait.

My point is, connoisseurship, for too long, has been about exclusion. The entire wine appreciation subculture was built around exclusivity and only recently have its participants been working to break away from that perception (to their credit, the younger generation of wine appreciators—Gen Xers and millennials—have been eschewing pretension and trying to make it more of the people's beverage). When the craft beer movement started to pick up steam, the most devoted aficionados (I count myself among them) were determined to prevent such highfalutin attitudes from seeping into the category. For the most part, they've been successful (though we all know a handful of jerks who still sip with the utmost snootiness). And since sake is a small (but growing) niche-within-a-niche category in the United States, those of us who have fallen in love with the beverage have the chance to keep the snobbery from seeping into the drink. I'm hoping this book will help the anti-snootery cause.

To that end, feel free to take my tasting notes with a grain of salt. I recognize the irony in saying such a thing in the introduction of this book, but again, when it comes to flavors and aromas, it's all incredibly subjective. Over the years, as I've studied to train my nose and palate to discern all sorts of nuances—or, truthfully, maybe two or three at a time—I'd been massively critical of myself when I couldn't pick out particular notes that I was "supposed to" (such as those in published reviews or supplier sell-sheets). I eventually accepted the fact that there are no right or wrong answers in tasting (regardless of what some ego-driven critics might say).

As a means of helping you "taste through text," included with all of my sake descriptions is a number known as the Sake Meter Value (SMV, an acronym you'll encounter throughout). The SMV indicates how sweet or dry a sake is (these numbers aren't mine; they're a measure of the sake's specific gravity—the density of the sake relative to water). Dry sakes will have a positive SMV; sweet ones will have a negative SMV. The higher the positive number, the drier the sake is. The more negative the number is (i.e., -10 is more negative than -5), the sweeter it is. The SMV is about as objective as sake tasting gets.

A word that's going to pop up quite a bit in my tasting notes, as well as many other places throughout this book, is umami. The term gets kicked around quite a bit in the culinary and beverage worlds (my own first encounter with it was in an introductory beer-tasting seminar). Most of us (myself included) have grown accustomed to just nodding along when someone mentions umami. But we have never quite grasped what it exactly is. Well, here's all you really need to know (and if you're already familiar with the concept, you have my permission to skip ahead).

Most of what we think we're tasting, we're actually smelling. Everything else is aroma. Our noses are tricking us into thinking we're tasting it (which is why you can barely taste anything when you have a cold even though your tongue is perfectly fine). Traditionally there really are only four basic flavors: sweet, sour, bitter, and salty. But a fifth has emerged, and that is umami—which, as you can tell, is of Japanese origin. The savory taste, which goes beyond basic saltiness, is the core characteristic of Japanese cuisine. Various types of fish, as well as mushrooms and seaweed (think the nori that holds your sushi rolls together), which are common in Japan's national diet, contain pronounced umami components. Sake has been brewed for centuries with that diet in mind. But that doesn't mean it's a strictly Japanese concept. Indeed, it's found in virtually every world cuisine. It's just that the Japanese were the first to name the flavor. Sake sommelier Jamie Graves likes to use the mouthwatering qualities of sundried tomatoes and parmesan cheese as examples when he's describing umami.

And umami is a flavor element, to varying degrees, in sake itself. I'll end the sensory lesson there because I've never worked professionally as a chef and I don't want to get letters.

5

Beyond recognizing sweetness or dryness or the vague presence of umami, think of your flavor impressions as an argument. If you can back it up intelligently, no one can really tell you you're wrong. Just make sure never to contradict yourself. You don't want to say the texture of a particular sake is "sharp" and "smooth." Pick one and stick with it. Other than that, go nuts. Don't let your tasting journey be tethered to the conventions established by the so-called "experts." Get to know sake on your own terms. The best place to start is with a little history lesson.

6

SAKE THROUGH HISTORY

I t's impossible to talk about sake without first addressing its place in Japanese history. In many ways, sake *is* Japanese history. The story of Japan is really a series of historical epochs that span millennia, which means it's easiest to explore the beverage in the context of those periods.

JOMON PERIOD
(14,000–300 BCE)

This is when pre-history morphs into actual history. The era is named for a pattern commonly found on the surfaces of earthenware vessels crafted during that roughly 140-century period. Artisans achieved the jomon—a rope-like texture—by pressing cords onto the wet clay.

Sake had yet to emerge in ancient Japan in the Jomon period. Instead, the drinking habits resembled those of the more Western parts of the world. The ancient Japanese were probably drinking wine fermented from whatever grapes grew in the wild back in those days (grape residue was found on the remnants of the era's clay pots). So whenever you read an article about how consumers in Japan are drinking more and more wine these days, remember that they were into it long before it was cool.

YAYOI PERIOD
(approximately 300 BCE–300 CE or 250 CE)

Named for a part of Tokyo where archaeological discoveries associated with the post-Jomon era were first discovered, the Yayoi period is believed to be when the Japanese started cultivating rice. The grain

7

likely made its way to Japan via China or the Korean peninsula, first arriving in the Japanese southern island of Kyushu and expanding through the rest of the country's main archipelago. The best evidence of the beverage's approximate time of origin surfaced in a Chinese historical text, *The Book of Wei*, which included the first documented mention of the Japanese rice-based drink commonly consumed in Japan.

The sake of that time was very crude and barely resembled the modern version. That proto-nihonshu was referred to as kuchikami-no, which translates to "mouth chew," a term that wasn't figurative. A very critical step in the early production process involved chewing and spitting rice into clay pots. "What is this madness?" you ask. People weren't trying to see how gross they could be; there was actual science involved—even though they didn't quite know it at the time.

Our own saliva contains the enzymes amylase and diastase, which break down starches and convert them to fermentable sugar. Yeast can't feast on complex starch molecules, so there needs to be some pre-fermentation step to make them palatable for those microorganisms. There are modern, less spit-centric ways to make that happen these days (which we'll get to later), but back then there really were no other options. And the salivary method was not unique to Japan. The process existed in the early, pre-Columbian civilizations of Central and South America. Instead of rice, the indigenous peoples were chewing and spitting corn to make a beer-like maize beverage that came to be known as chicha de jora. Alcohol bases that contain more starch than sugar—particularly grains and potatoes—require those sugar-making enzymes. That's why, for instance, barley used for beer and whiskey requires malting. The malting process involves soaking the grains in water to allow those cereals to germinate and then drying them to halt germination, resulting in the necessary fermentable sugars. Alcohols made from sugary fruits—like wine—don't require that starch conversion.

— KOFUN PERIOD —
(approximately 250–538 CE)

The time periods start getting a little more compressed at this point, as this is pretty much the dawn of recorded history in Japan. This era is named for the burial mounds in which rulers and the wealthy were entombed. These mounds were often shaped into a keyhole-like pattern (among other less common shapes). Sake's role in religion was particularly pronounced during the Kofun period. Typically it was used as an offering to the gods to ensure a bountiful rice harvest. It was also customary for the common folk to present sake to the imperial family.

The nihonshu of the time wasn't the clear, crisp libation you're likely to find in stores, bars, and restaurants today. Known as doburoku, the sake had a milk-like opacity, as the rice solids aren't pressed out. (You might have encountered the cloudy sake style, nigori, and wondered how doburoku differs from that. Doburoku is sake in its pre-pressed form. *Nigori* is pressed and coarsely filtered, allowing some of the solid lees to get through. Nigori means "unclear" or "murky," and sometimes is incorrectly referred to as "unfiltered" sake. Filtration does occur.)

Present-day sake enthusiasts keep the Kofun period (and its signature drink) alive with the annual Doburoku Festival. And yes, fest-goers continue to offer the beverage to the ancient deities.

ASUKA PERIOD
(538-710 CE)

Historians argue over the exact delineation of the Kofun and Asuka periods, as there's a bit of an overlap in some of the traditions of the two eras. Asuka was named for the region just outside of what is today the city of Nara (near Kyoto on the large, central island of Honshu). During this period there were massive social, artistic, and political shifts, most notably through China's introduction of Buddhism to Japan. (Historians point to a window between 538 and 552 CE as the beginning of Japanese Buddhism, hence the year's common use as the demarcation point between Kofun and Asuka. However, some date the dawn of the Asuka period much later in that century.) One aspect of everyday Japanese life that didn't change that much was the production and consumption of doburoku sake.

NARA PERIOD
(710-794 CE)

The Nara period, on the other hand, was a time of great transformation in sake making. It was during the eighth century that the microorganism koji made its first appearance in nihonshu production. We'll get into the details about koji a little later, but it's basically the mold that rendered the kuchikami-no (the "chew and spit one") obsolete. The period takes its name from the city of Nara, at the time called Heijo-kyo, which became the Japanese capital for most of the eighth century. It was the dawn of what would become modern sake making.

——— HEIAN PERIOD ———
(794–1185)

The Heian period officially began when Emperor Kammu moved the Japanese capital from Nara to Heian-kyo—today known as Kyoto, a much more familiar name to those of us who live outside the country (and quite a beautiful city in its own right, with more than its fair share of stellar drinking establishments in which to enjoy a few bottles of fine nihonshu).

The medieval age in Japan was not unlike the same era across Europe in that monasteries were the big alcohol producers. But in this case it wasn't Roman Catholic monks making beer, but Buddhists brewing sake. (We really owe a debt of gratitude to monks the world over for the tasty delights we enjoy today!) The monks were making what is called souboshu, which simply translates to "monk's sake."

Over the course of those four centuries, sake's presence in society shifted from a beverage for the gods to one consumed by the general citizenry. Additionally, brewers took a giant leap toward modern sake-making techniques when they started using polished rice in their recipes—producing primitive versions of what, centuries later, would evolve into ginjo- and daiginjo-grade sakes. I'll detail those styles and their twentieth- and twenty-first-century production methods in greater detail in upcoming chapters. The shorter version is that sake brewers craft different grades of sake based on how much of the rice kernel is polished or milled away. With ginjo-grade sake, at least 40 percent of the grain is polished away. Daiginjo grade takes it further, with at least 50 percent of the kernel milled away. All rice starts as brown rice. The outer husk is very protein-rich, and the objective is to get closer to the starchy center. During the Heian period, this was accomplished with far less precision than it is today, when modern milling technology does the heavy lifting. Ginjo- and daiginjo-grade sakes have more delicate flavors than those with more of the grain remaining.

— KAMAKURA PERIOD —
(1192-1333)

Named for the Kamakura shogunate, the feudal government that ruled for nearly 150 years, the Kamakura period was when the country started to resemble the Japan of an Akira Kurosawa movie. The military shogun model of government emerged after the Minamoto clan defeated the Taira clan in the five-year Genpei War. In 1192, about seven years after the conflict ended, Minamoto no Yoritomo established his shogunate.

The period saw two major developments in the sake world—one rather dubious—which shaped its future. Sake production became much more commercialized, with a number of small breweries popping up and selling their wares to the populace. However, we all know the recurring theme throughout history, across continents, cultures, and beverage styles. Whenever there's a burgeoning booze industry without any sort of legislative or regulatory oversight, it's very likely that things are going to get out of hand (think eighteenth-century London's gin craze, for instance). Greater access to sake led to overconsumption among Kamakura-era citizens. It's not like there were any "Drink Responsibly" ads hanging up all over town. Since it was all new to them, nobody really knew how much was "too much."

11

The adverse social impact of the beverage led to a thirteenth-century version of prohibition in 1252 (nearly 700 years before America's 18th Amendment—always the pioneers). The temperance laws remained in place until the end of the Kamakura period, and sake industry growth stopped in its tracks for a while.

— MUROMACHI PERIOD —
(Approximately 1334-1467/1573)

The prohibition ends! Not only did the government allow sake business to boom once again during the Muromachi period—named for the Muromachi shogunate's ascension to power—the rulers also discovered alcohol's great potential as a source of revenue through taxation. Essentially, the anti-sake climate of the Kamakura period made way for a much more pro-alcohol environment. More sake producers meant more money for the government to do whatever it is the government does. By 1425, the Kyoto Prefecture was home to nearly 350 sake breweries. The region is still a hotbed of production today, despite a decline in general consumption of the beverage—thanks to the rise of other alcohols like beer, shochu, and whiskey.

—— SENGOKU AND AZUCHI-MOMOYAMA PERIODS ——
(1467/1573-1603)

Technically, Sengoku begins within what's traditionally defined as the Muromachi era, beginning in 1467 and overlapping with the Azuchi-Momoyama age (which begins around 1573 and is often considered by historians to be the tail end of the Sengoku period). Its defining characteristic was ongoing warfare, with feudal lords known as daimyo (ranked just below the ruling shoguns), constantly duking it out to try to expand their power (at the expense of the others, of course).

There was a great deal of beverage-related upheaval during the Sengoku/Azuchi-Momoyama era. During the many armed battles of the time, daimyo had been known to destroy Buddhist temples, which led to a dramatic reduction in the volume of souboshu monks had been producing.

The age also saw waves of Western missionaries arriving in Japan, trying to spread their European spiritual and commercial philosophies. Among those was Francis Xavier, who is said to have brought wine to the daimyo. There's a good chance that it was the first time the Japanese imported any alcohol from the West.

It was also around this time that Japanese distillation began in earnest, introducing the spirit shochu. The southwestern home island of Kyushu was responsible for most of the country's distilling activity (as it is today—about 90 percent of all shochu volume comes from Kyushu), and it gradually spread through the rest of Japan.

—— EDO PERIOD ——
(1603-1868)

As the wars started to wind down and the Tokugawa shogunate ruled Japan from the city of Edo (now modern-day Tokyo), the country moved further and further into the modern world—which meant far more advanced methods of sake production became prevalent over the course of the roughly two and a half centuries of progress.

12

In the nineteenth century, pasteurization—or "hi-ire"—became a critical component of sake making. (Even though Louis Pasteur discovered the microbe-killing liquid heating practice that bears his name in 1865, some sake producers reportedly were performing a similar, unnamed technique several hundred years prior. However, before Pasteur, no one really knew that bacteria were being killed; they just knew that heating kept the sake from spoiling. After Pasteur defined the process, it became the norm in nihonshu making.)

As brewers gained a greater understanding of the role microbiology played in alcohol making, they discovered that the yeast starter, known as shubo—composed of steamed rice, water, and the beneficial mold known as koji —works best in an acidic environment. Typically, producers add lactic acid to kill off unwanted bacteria while the yeast propagates (yeast can survive alongside a certain level of lactic acid). Too much water, koji, and steamed rice at one time can dilute the lactic acid–rich environment, however, opening a window for unwanted microbes. So instead of one addition, the brewers add the water-koji-steamed rice combo three times over the course of four days.

The Edo period is also when clear sake emerged. Coinciding with the emergence of translucent, lightly golden European beer in the 1840s, sake brewers added wood ash to the beverage to clarify the beverage. Today, that practice has evolved to become what the sake industry calls the activated carbon filtration method.

Sake makers during this period also started to recognize the importance of the water source and discovered that some waters were far superior to others in beverage production. We'll explore the role of springs, melted snow, and other sources in the upcoming sections on sake ingredients.

—— MEIJI PERIOD ——
(1868–1912)

Those who have seen the Tom Cruise movie *The Last Samurai* have some familiarity with the Meiji period, or at least Hollywood's version of the era. Cruise's character, a US Civil War, vet, heads to 1870s Japan to advise the people on modern Western warfare and then the movie eventually follows the same formula as *Dances with Wolves*.

The era began with the Meiji Restoration, during which all political power for the entire country was consolidated under Japan's emperor

Meiji. It was essentially the end of the Japanese feudal system and the beginning of Japanese prominence on the international stage. It also marked the time that the most of rest of the world got its first taste of sake.

In 1873, nihonshu had its coming-out party of sorts at the Vienna International Exposition. It was the first World's Fair in which Japan had participated. Before that, Japan exported a few drops of nihonshu, mostly to Southeast Asia.

At the turn of the twentieth century, Japanese brewers bottled the beverage for the first time; prior to that it had been sold in large vessels. The standard container became the 1.8 liter glass bottle, which is still widely used. You're more likely to encounter a 720-ml bottle in stores and most restaurants, but if you visit an izakaya and order nihonshu by the glass, there's a good chance your server will be pouring from one of these gargantuan magnum containers.

⸺ TAISHO PERIOD ⸺
(1912-1926)

In the fourteen and a half years that Emperor Taisho ruled Japan, the country took a brief detour into liberal democracy (*Taisho* loosely translates to "great righteousness"). In a sense, it was the calm before the storm that would be the early years of the subsequent Showa period. The biggest sake development during this time was the move from storing the beverage in wooden containers to enamel-encrusted iron tanks. From a flavor standpoint, this was monumental. Anyone who's ever consumed a barrel-aged brandy knows that wood influences the flavor of the liquid it contains. It also absorbs some of the beverage through its pores. Enamel tanks, however, do neither.

⸺ THE SHOWA PERIOD ⸺
(1926-1989)

In the first decade-plus of the reign of Emperor Hirohito (post-humously renamed the 124th Emperor Showa), Japan shifted from "righteous" liberalism in favor of militaristic totalitarianism. As the country's aggressive expansionism culminated with World War II, sake evolution mostly halted, save for the creation of a classification system that would remain in effect in some form for decades to come. In 1943, the government implemented the kyubetsu-seido, which differentiated sakes by four classes (the system would be revised in 1962, reducing the number from four to three). For instance, a brewer paid higher taxes for "first class" sake than for "second class," and so on. Naturally, many producers figured out how to game the system. They'd brew a sub-par sake and voluntarily pay a higher tax rate so their beverage could gain a more prestigious classification. So you could be a humble artisan with limited resources making world-class stuff but lacking the funds to afford higher classification. Meanwhile, your competitor down the street, a massive industrial operation flush with cash, would essentially bribe the government to make its product appear more premium. It was a practice that was doomed to fail.

During and immediately after the War, Japan experienced a massive rice shortage. Producers resorted to adding distilled spirits to sake to compensate for the diminished ingredient supply and boost production. Brewers had been using similar practices for centuries, adding shochu to their nihonshu to prevent spoilage. Many continue to do so today with some offerings—not for lack of resources or to cut corners, but to accentuate a specific flavor profile with the distilled spirit. It's easy to distinguish which products have added spirit and which do not—a sake labeled as junmai, meaning "pure," contains only alcohol that results from the rice fermentation process.

15

—— HEISEI PERIOD ——
(1989-present)

Emperor Hirohito/Showa died in January 1989 and his son, Emperor Akihito, acceded to the throne (as per tradition, he will be renamed Emperor Heisei after his death). That same year, the government again revised the sake classification system, reducing the number of classes from three to two, ikkyuu (first class) and nikyuu (second class). It was something of a last gasp for the kyabetsu-seido, and in 1992, the system was eliminated for good. Taxation is now based on alcohol content, and classification is tied to something else entirely: seimaibuai. Producers mill away a certain percentage of the outer rice kernel to achieve a desired profile. The percentage of rice that remains is its seimaibuai. So the lower the seimaibuai, the higher the grade.

I once had a conversation with a food critic who asked me why wine writers can't seem to fully get their heads around sake. I told him it wasn't the fault of the writers but a grading system that's only as old as MTV's *The Real World*. Ginjo- and daiginjo weren't legal sake classifications before the early '90s, though the polishing practices and the terms existed for decades before that.

You'll recall that with ginjo and daiginjo sake, at least 40 and 50 percent of the grain is milled away, respectively. So if, say, 42 percent of the outer kernel is polished off, it would be used in a ginjo with a seimaibuai of 58. If 60 percent of the rice is polished away, the resulting daiginjo would have a seimaibuai of 40 percent.

Daiginjo tends to be the pricier grade, mainly because brewers must use far more rice to make it since the mass of each kernel has mostly vanished in the milling process. Some producers have been known to achieve a seimaibuai in the single digits for a daiginjo release, but those are extremely rare. But you can imagine that if a sake has a seimaibuai of 9 percent, it takes a whole lot of rice to fill that bottle.

Ginjo and daiginjo—especially the latter—typically have a more delicate flavor and are served chilled to fully showcase their nuances. But I'll go deeper into proper serving temperatures (and there are many!) in a later chapter.

16

Now that we're caught up on a couple millennia worth of Japanese alcohol-making history, let's take a closer look at sake ingredients.

CHAPTER 2.

THE BUILDING BLOCKS

s far as composition, sake is pretty darned simple. Its components are basically rice, water, yeast, koji, and in some cases, distilled alcohol. Let's have a look at each of those.

——— RICE ———

A common misconception about sake is that it's essentially "rice wine." Sake's alcohol content tends to be fairly close to that of wine, but its production process is much closer to that of beer. Plus, both beverages are made from grains, and sake makers are called brewers, not vintners.

Well then, you ask, isn't whiskey made from grain? Indeed it is, but whiskey is distilled and sake is not. One of my pet peeves is when writers for various media outlets refer to sake as a "spirit" or call its production facility a "distillery." I read an article that did that barely two days before I wrote that line. Fortunately, I was saved from my compulsion to be a pedantic ass in the comments section, as someone else had already beaten me to it.

Before getting into the varietals used specifically for sake making, I want to quickly distinguish the three broad categories of rice grown around the world. There's Indica, the long-grain variety, which is the most ubiquitous because it commands about an 80 percent share of the global rice market. It's cultivated in India (of course), Thailand, Vietnam, China, and the United States and has a dry, loose consistency.

Medium-grain Javanica rice—typically found in Southeast Asia, Spain, and Italy—is far less prominent, barely making a mark in the worldwide market share. It's a bit stickier than Indica, but not as sticky as the third variety.

That third variety, Japonica rice, includes the 100 or so breeds used in sake making. About 90 percent of it, however, is table rice. Despite its name and common uses, Japonica didn't actually originate in Japan. New evidence suggests that it actually originated in Southeast Asia, eventually traveling through Yunnan and then through the lower Yangtze River region before finally making its way over the East China Sea to Japan.

The version of Japonica rice used to produce most sake differs from the variety you're likely to eat as part of a meal. While it's perfectly possible to brew sake with common table rice, the vast majority is made with what's known as shuzo-kouteki-mai (but calling it "sake rice" is perfectly acceptable).

shuzo-kouteki-mai is considerably more expensive than table rice. Since it's a lot harder to cultivate, far less of it is produced, so producers don't get the same bulk discounts they'd get for the more edible stuff. A kilo of table rice costs an average of 300 yen ($3, give or take) in Japan, while sake rice costs about double that.

Only about 1 percent of the total rice grown in Japan is shuzok-outekimai, though 5 percent of the country's rice is used to make sake. In other words, 4 percent of the nation's table rice crop finds its way into a nihonshu brewery (to be used for the lesser-quality products).

So what makes "sake rice" so special? First, the kernels are much larger than those of the table variety, yet they are actually very fragile. They're also like sponges in that they have an incredible capacity for water absorption.

But there are two characteristics in particular that make sake rice so desirable: shinpaku and gaikonainan. Shinpaku, or "white heart," means the kernel has a dense concentration of starch in its core. The outer part of the kernel tends to be more amber-brownish, and the deeper you get to its center, the more opaque, strikingly white starch you'll encounter. While it has an abundance of starch, shinpaku has very little protein and fat—too much of which can have some undesirable effects on the desired flavor of sake.

Gaiko Nainan comes into play after the rice is steamed. The term means firm on the outside and tender on the inside or, to borrow an oft-used foodie phrase from another culture, al dente.

Even within the universe of shuzokoutekimai, not all rice is created equal. There are four significant varieties of premium sake rice that get quite a bit of attention. The first is Yamada Nishiki (sometimes written as Yamadnishiki), developed in 1923 and officially named in 1936. Those who intend to produce a nihonshu that's complex, aromatic, and fruity often opt for Yamada Nishiki. If they're looking for a light, clean sake, they might try Gohyaku Mangoku, introduced in 1938 and earning its moniker in 1957. Gohyaku Mangoku means "5 million koku"— koku is a unit of rice measurement—honoring an important milestone in grain-growing circles: 1957 was the year that the Niigata Prefecture, practically the Napa Valley of sake, first achieved a 5 million koku crop yield.

Discovered as far back as 1859, Omachi is considered an "heirloom rice," as the species grown today is still from that original strain. Grown almost exclusively in the Okayama Prefecture, Omachi is notoriously difficult to cultivate—partly due to the fact that the plant grows rather tall and is very susceptible to high winds and inclement weather. It's also known to be quite temperamental during the brewing process and requires extra care. But the deep, complex, "earthy" flavor it imparts to sake has made it worth the trouble for many brewers. Many assert it's the best variety of rice to pair with food, given its robust characteristics. Keep an eye on this one. If the word *omachi* shows up on a bottle's label, you'll know you're in for a treat.

An easy way to remember the next variety, Miyama Nishiki, is to think of winter. The breed has built a reputation on its tolerance to the cold, and the sake it produces tends to be brisk and sharp, like winter itself. Plus, Miyama Nishiki grows primarily in the Nagano Prefecture, which, if you recall, hosted the 1998 Winter Olympics.

Since I mentioned snow sports, I'd be remiss if I didn't talk about Niigata Prefecture, one of the big, snowy ski areas in Japan, located just north of Nagano and the birthplace of the rice variety, koshitanrei. A joint team of Niigata Prefectural Sake Research Institute and Niigata Agricultural Research Institute scientists introduced koshitanrei when they cross-bred Gohyakumangoku and Yamada Nishiki. Koshitanrei inherited the best traits from both of its parents—the smooth, clean, dry characteristics of the former and the highly fragrant and soft

19

elements of the latter. It's also quite the sturdy grain that's easier to polish and very hospitable to koji spores, therefore easing the rice-koji-making process.

Travel a bit northeast of Niigata and you'll arrive in Yamagata Prefecture, famous for the type of rice known as Dewasansan (a word I love to say because it rolls off the tongue like poetry). Dewasansan yields a mildly sweet-to-medium dry sake with a light-to-moderate aroma, depending on the polish ratio.

If you head back down to the Kyoto Prefecture and happen to see the word *Iwai* on the bottle, you're in for as local an experience as things get there. Iwai is Kyoto's pride and joy rice variety, first cultivated in the early 1930s. It earned the distinction of "recommended" variety for sake production, but following World War II, food shortages forced farmers to allocate more acreage to table rice. Iwai made a comeback in the 1950s, only to fall out of favor once again in the early 1960s. But its most recent resurgence began around 1992, when Kyoto brewers were keen to develop a truly local sake, using only ingredients native to the prefecture. Iwai is known for the subtle, delicately sweet aroma and mild creaminess it imparts to sake.

The last variety, Hattan Nishiki, is comparatively brand new. Developed in 1973 and commercialized in 1984, Hattan Nishiki's claim to fame is its relative resilience. It grows shorter than other varieties, which is advantageous when cultivating the grain. The shorter the plant, the less vulnerable it is to heavy winds. It's also fairly disease-resistant, which means it doesn't require much helicopter parenting.

There are more than 90 other rice varieties commonly used in making sake, but I just wanted to detail the kinds you're most likely to encounter as you sip your way through the nihonshu realm.

RICE GROWING AND HARVESTING

The active growing season for rice in Japan is about six months, beginning in March and culminating in September. Japanese farmers employ a process known as wet rice cultivation—versus upland cultivation, which is used throughout much of the rest of Asia—which, as you'd expect, involves planting and growing in wetland areas.

In the spring months, March through May, farmers sow the seeds that eventually will become the starchy grain. Then, in June, the growers start digging shallow trenches in the rice paddies to ensure proper irrigation and even water distribution, a process known as mizukiri. That's followed by the mid-summer practice called nakabashi, whereby the farmers drain the water from the paddies. It's ready for harvest in September, just before the official start of the traditional sake brewing season.

── HOLY WATERS ──

Contrary to what may seem likely at this point, rice is not the most important ingredient in nihonshu. Every ingredient is of equal importance. But there's always one that tends to be taken for granted in sake—and in most other alcohol beverages—and that ingredient is water.

Beverage companies have for quite a long time—probably since the dawn of modern advertising—trumpeted the quality of the water used in their products. How many times has Coors hit us over the head with the fact that the brand uses pure, Rocky Mountain spring water? But in reality that tends to be mostly a marketing ploy.

21

In Japan, however, that is not the case. The type of water sake brewers use is every bit the celebrity that the base grain is. Brewers aren't just turning on the tap, filling a bucket, and hoping for the best. Before they produce a single drop, they decide whether the desired flavor profile calls for either hard or soft water.

I had sometimes heard people complain about the hardness or softness of water when washing dishes or trying to rinse shampoo out of their hair, but I didn't know the scientific definition, or its relevance to drinking, until I took a class with the Sake School of America. The degree of hardness correlates with the concentration of calcium and magnesium in the sample. There's a calculation involving milligrams per liter of each of those elements, but no one opened this book because they wanted to do math. To make a long equation short, the resulting level of calcium carbonate determines whether water is hard, soft, or somewhere in between. A sample with zero to 59.99 parts per million (ppm) of calcium carbonate is considered "soft." One containing between 60 and 119.99 ppm is defined as "semi-hard." The threshold for "hard" is 120 ppm, and that extends to 179.99 ppm. Anything 180 ppm and above is considered—get ready for this ultra-scientific term—"very hard." A ppm is equivalent to 1 milligram of calcium carbonate in a liter of water.

Here's the part that can get a bit eye-roll-inducing, especially if you take it at face value. There's a distinction between "masculine" and "feminine" sake, and you can probably guess what those terms mean in this context. Hard water is best for "masculine" or "strong" (in flavor and mouthfeel, not alcohol content) nihonshu, while soft water is favored for light and clean, "feminine" sake. I don't think any sexism was intended when sake producers came up with this terminology, but it still makes me wince a little.

Okay, I will dismount my high horse.

Regionality plays a major role in the hardness and softness of water, which is why sake produced in certain areas can become known for a distinct flavor. This is also why some local water sources are prized well above others. One of those is Miyamizu, which originates in the Nada region within the Hyogo Prefecture. Miyamizu first gained prominence around the nineteenth century when a well-known sake brewer of the era, Tazaemon Yamamura, noticed how much different the nihonshu produced in the region tasted versus those made elsewhere. The water gets its signature characteristics from a high concentration of phosphorus and potassium. Those nutrients are known to feed the koji mold, which is also a big plus.

Though it's rich in minerals, one thing that it fortunately lacks is iron. Yes, you should be eating your spinach to get your iron, but you do not want to find any of it in sake. It has a very off-putting impact on flavor.

The sake made with it tends to be dry and—brace yourselves—"masculine," thanks to its high minerality.

In the Kyoto Prefecture, also on Honshu, the Fushimi region is home to its own fine water source. About 1,000 years ago, the locals happened upon a spring producing fairly aromatic water, which they ultimately named Gokosui, aka "honorable aromatic water." Thanks to that quality of the water, Fushimi-made sake is known to be sweeter.

Mount Fuji is responsible for another noteworthy water classification, Fukuryusui. Snow and rain seep deep into the mountain's volcanic soil, eventually becoming extremely pure spring water, while volcanic rocks act as a natural filtration agent. The sakes of the region can be quite crisp with a soft mouthfeel, all thanks to a volcano.

But for as acclaimed as these sources are, you can't really go wrong with any of the water in Japan. The country as a whole gets an average of 60 inches of rain a year, plus a large amount of snow in higher elevations and on the northern home island of Hokkaido. The precipitation eventually sinks into the ground and feeds the wells that, in turn, feed the sake industry.

⏤ WEE BEASTIES ⏤

We're not done with the science stuff quite yet—it's time to talk about microbiology. And since these microbes are the bringers of alcohol, that makes them your new best friends.

The one microbial element that sake shares with virtually every other alcohol beverage is yeast. And as many of you beer and whiskey drinkers probably already know, yeast—typically of the species *Saccharomyces cerevisiae*—eats sugar and converts it to alcohol and carbon dioxide. Naturally carbonated beers—those that don't involve the addition of CO_2, aka "forced carbonation"—get their bubbles from a secondary fermentation.

But one microbe sake doesn't share with fellow fermented and distilled beverages (with the exception of the Japanese spirit shochu) is a fungus known as koji. The mold is a critical component in the sake-making process as it produces enzymes that break down rice's starch into fermentable sugar.

Koji isn't a necessary part of beer production, for instance, as beer brewers usually use malted barley—grain that's been allowed to germinate by soaking in water and then dried. The malting process produces the necessary enzymes. The malted grain is then milled and again combined with water—a process called mashing—which enables the enzymes to go to town on the starch.

Sake brewers inoculate koji onto steamed rice in a koji muro (koji room), which is kept at a sweltering 95 degrees Fahrenheit (and we're not talking dry heat here) for about 48 hours. And the production team can't just set a thermostat, spread out the rice, and walk away for two days. The process involves nonstop TLC (I'll go more in-depth on that process in chapter 6).

Once the process is complete, koji leaves its signature all over the finished sake, playing a central role in the flavor and aroma of the beverage.

There are three main categories of koji microbes, all delineated by the color of their spores. There's white koji (shiro-koji), which is commonly used to produce the aforementioned spirit, shochu. It's known for producing a considerable degree of citric acid and powerful starch-converting enzymes. Shiro-koji, scientifically known as *Aspergillus kawachi*, is actually a mutated form of black koji (kuro-koji) and is the favored microbe used to produce the Okinawan spirit, awamori (essentially Okinawa's answer to shochu), hence its scientific name: *Aspergillus awamori*. Many shochu distillers also use black koji, which tends to give the beverage earthy, almost "dirty" qualities.

That brings us to yellow koji (ki-koji), better known among microbiology types as *Aspergillus oryzae*. This is the variety that's used to make sake, as well as non-alcohol products like soy sauce and miso. Shochu producers also offer yellow koji versions of their spirits, which one often can discern from their fruity, citrusy aromas. Rice-based shochus made with yellow koji frequently are the most sake-like of the spirit.

Once the sake makers inoculate the rice with ki-koji and finish the entire two-day process in that sauna of a koji muro, the rice koji that emerges can be classified within one of two categories. Souhazegata is the most aggressive of the rice koji, as the fungus covers the entire surface of the kernel and penetrates deep into its center. The other is tsukihazegata, where the fungus still bores its way to the core of the kernel but only sparsely covers the surface. Souhazegata tends to yield more intensely flavored sake, while tsukihazegata is the favored rice koji for cleaner, more delicate styles like ginjo or daiginjo.

That's about as deeply into koji as I'm planning to go, and I promise things won't be so technical going forward.

At this point, the microbial process starts to look a lot like that of beer and other fermented beverages. The starch has been broken down into glucose, and it's time for the yeast to take over. Though some nihonshu breweries use ambient yeasts that naturally inhabit their facilities, most are buying *Saccharomyces cerevisiae*—which is also used in beer production—from the Brewing Society of Japan.

That's not to say sake brewers are using the exact same strains that beer makers do. Over the years, Japanese fermentation researchers have developed variations that work especially well with the country's rice-based adult beverage. Since sake fermentation takes place in the presence of lactic acid, the microbial strains must be particularly adept at doing their work in acidic environments. Brewers and researchers across Japan have cultivated multiple yeasts optimized for different flavor profiles and sake grades.

➤➤ KYOKAI NO. 7: Commonly known as K7, this is the strain most commonly used in sake production. Its point of origin—or, rather, the place where brewers first isolated it—was the Miyasaka Brewery in the Nagano Prefecture back in 1946. K7 is popular for its stability and subtle aromas, and is usually the strain of choice for junmai and honjozo.

➤➤ KYOKAI NO. 9: It's known as K9, but it's certainly no dog. First isolated at Koro Brewery and cultivated at the Institute for Kumamoto Brewing in 1968, K9 is an ideal option for ginjo and daiginjo. It's known to impart floral, fruity aromas—the signature notes of those highly polished grades.

➤➤ KYOKAI NO. 12: Often known as "Urakasumi yeast" after its point of origin, Urakasumi Shuzo in Miyagi Prefecture, this strain actually predates No. 9 by about two years. The aroma it produces is fairly mellow, and it's known to perform well in lower temperatures—not a bad quality to have in a prefecture in the northern quarter of Honshu.

➤➤ KYOKAI NO. 14: Having been developed in 1995 by the Brewing Society of Japan, No. 14 is the relative baby among yeast strains. The aromas it produces are floral and fruity (so another good candidate for ginjo), but its acids are on the low side.

➤➤ KYOKAI NO. 701 AND KYOKAI NO. 901: Respectively, these two strains are mutated forms of K7 and K9, exhibiting similar characteristics of each. However, they produce far less foam than their predecessors. The Research Institute of Brewing developed K701 in 1969 and K901 in 1975.

OTHER NOTEWORTHY STRAINS

• **Alps Yeast:** This low-acid strain was developed in Nagano Prefecture in 1991, imparting flowery and fruity aromas.

• **F710:** This strain is another low-acid variety developed the same year as the Alps yeast in Fukushima Prefecture, but with a more subdued aroma.

• **Akitaryu-hanakoubo:** Also known as AK-1, this low-acid strain developed in Akita Prefecture in 1981 is known to produce floral and fruity aromas.

I truly believe that microbes—be they yeast, bacteria, or koji—will be the next frontier in sake innovation, similar to how certain microorganisms have inspired experimentation among craft beer brewers. If nothing else, such experimentation enhance a brand story. And as international markets develop for sake—I'm looking at you, North America—hopefully the ensuing geekery that emerges will increase demand for such microbial innovation.

CHAPTER 3
TASTING NOTES: ICONOCLASTIC RICE

E ach of the sakes we sample below highlights a single type of rice, some of which are heirloom varieties. This will highlight the role that the grain plays in flavor and aroma.

BIZEN OMACHI JUNMAI DAIGINJO

Brewer: Tamanohikari Shuzo
Prefecture: Kyoto
SMV: +3.5

Tamanohikari uses only bizen-omachi, a rare rice variety considered to be among the best, to craft its junmai daiginjo. The nose reveals a little bit of fruit, but it's mostly restrained. There's a dry, sharp hit that lingers when you first sip it, but a slight sweetness on the finish.

SEIKYO OMACHI JUNMAI GINJO

Brewer: Nakao Shuzo
Prefecture: Hiroshima
SMV: +3.0

"Crisp and bright" immediately come to mind with this junmai ginjo brewed with Omachi rice. After the first few sips, it was a bit difficult to pin down. There's a definite sharp, earthy minerality and, dare I say, a vague dustiness in its texture. Mostly, though, it settles into familiar, fruit-forward junmai ginjo territory—mostly melon, but maybe a little apple as well. When I first tried this one at Yama Sushi in Portland,

Oregon, the menu described it as "possibly the best sake available in the United States." That might be overselling it a little bit, but it's definitely worth a few rounds.

WATARI BUNE "55"

Brewer: Huchu Homare Shuzo
Prefecture: Ibaraki
SMV: +3

This is a real treat for those looking to acquaint themselves with sake made from heirloom rice. In this case, it's the Watari Bune variety, which is the granddaddy of Yamada Nishiki. The "55"—its polish ratio—is a little on the sweet side up front, but it finishes much drier. Its fruit notes tilt mostly toward melon, but you wouldn't be wrong if you detected a little green apple in there as well as some very subtle pineapple and even licorice.

TAITEN SHIRAGIKU JUNMAI MIKINISHIKI

Brewer: Shiragiku Shuzo
Prefecture: Okayama
SMV: +3

I'm intrigued by the way this heirloom rice sake—in this case the elusive Mikinishiki variety—teeter-totters between sweet and tart before finishing in contrastingly dry territory. That dynamic evokes flavors of sour cherry, perhaps black currant, and more than a sprinkling of citrus. The mouthfeel is remarkably heavy, but in the best possible way. It almost makes me want to savor it longer.

KUROUSHI ["BLACK BULL"] OMACHI

Brewer: Nate Shuzoten
Prefecture: Wakayama
SMV: +3

Balance. That's really all I have to say. Seriously, I was struck by how much Kuroushi's heirloom rice sake is equal parts tropical fruit basket and bowl of steaming rice. The texture can be a bit sharp at times, but that just makes this relatively easy drinker even more interesting.

TOMIO IWAI JUNMAI GINJO

Brewer: Kitagawa Honke
Prefecture: Kyoto
SMV: +2

The unexpected flavor complexity is off the charts—in addition to some standard cooked-rice quality (from its namesake rice variety), there's an intriguing sweet-and-sour element. It's distinctly vinegar-like, along the lines of a pickled plum or some other stone fruit. I tasted this for the first time at Abu Racho, a bar and bottle shop in Kyoto that offers tastings of some 80 different sakes from the prefecture, and had to buy a bottle.

JUNMAI KARAKU IWAI

Brewer: Shoutoku Shuzo
Prefecture: Kyoto

Iwai helps deepen the character of this Kyoto junmai-shu. There's a berry and cherry aroma that's coupled with a bit of baking spice like cinnamon and clove (so you could say a fruit pie aroma). A neat little iconoclast of a sake, if there ever was one.

29

EIKUN BIG HAWK

Brewer: Saito Shuzo
Prefecture: Kyoto
SMV: +3

Eikun Big Hawk junmai ginjo is a celebration of Kyoto in a glass. It uses the prefecture's highly sought-after Iwai variety and the Fushimi district's incomparable water in this creamy, sweet-rice-forward affair with a little bit of melon thrown in to keep everyone engaged. If you were to dip a slice of honeydew in a bowl of rice, you might get a similar experience. There's an underlying sweetness up front, but that gets clipped pretty fast by a dry, somewhat sharp finish.

KOZAEMON TOKUBETSU JUNMAI SHINANO-MIYAMANISHIKI

Brewer: Nakashima Shuzo
Prefecture: Gifu
SMV: +6

Miyama Nishiki rice, milled to 55 percent, is the marquee attraction in the dry, yet balanced, tokubetsu junmai. It's as smooth as polished jade and quite fruit-centric—I'd say apples and similar orchard fruit—on the nose.

HAKUSHIKA JUNMAI YAMADANISHIKI

Brewer: Tatsuma-Honke Shuzo
Prefecture: Hyogo
SMV: +1

There's a certain dessert-y element to this junmai that uses the famous Yamada Nishiki rice. I don't mean it should be consumed with dessert, but that it has a few confectionery things happening on the nose. Sometimes there's a sheet-cake, buttercream-icing vibe. At other times it's more reminiscent of yogurt, tapioca pudding, or mochi. There's also a little bit of citrus on the palate, with an appreciable acidity.

HAKUTSURU SHO-UNE

Brewer: Hakutsuru Shuzo
Prefecture: Hyogo
SMV: +2

Another Yamada Nishiki-centric offering, Sho-Une combines the most celebrated sake rice variety and the legendary Miyamizu water from Nada for a sake that's a good introduction for junmai daiginjo newcomers. The aroma is a bit more subtle than many of its ilk, but you'll definitely detect some orchard-fruit notes (I usually get pear, but I know others who swear there are hints of peach).

MINAKATA JUNMAI YAMADA NISHIKI

Brewer: Sekai Itto
Prefecture: Wakayama
SMV: +5

The restrained aroma of this variety can be deceiving. A few sips in and you'll discover notes of cedar and a touch of earthiness. It's pleasantly dry with a somewhat sharp texture.

·CHAPTER 4·
GRADES, STYLES, FLAVORS, AND AROMAS

I've been throwing a lot of multisyllabic, hard-to-pronounce stylistic terms around, and it's high time that I start explaining what some of those are.

One of the reasons sake seems unapproachable to the average American consumer is the fact that there are grades, styles, styles within grades, grades within styles, flavor categories, and a whole host of other complex concepts. Additionally, it's often difficult for a US drinker to distinguish one bottle from another on a store shelf because, to average Western eyes, the labels are difficult to decipher.

Let's get the low-hanging fruit out of the way first.

FUTSU-SHU

Futsu-shu ("ordinary sake") is not a style per se, but an umbrella term used to describe "value," sub-premium sake. Think of it as the Busch, Keystone, or Natural Light, or in oenological terms, the table wine of the nihonshu universe. They're made inexpensively in big, industrial production facilities, but they're by no means "bad." Quite the contrary, many are actually amazing and produced in an artisanal manner (though most are associated with big, industrial production). Some even satisfy the criteria for honjozo classification, but the breweries won't label them as such to avoid alienating the everyday drinkers (some of whom find the premium classifications highfalutin). Like sub-premium beer and wines priced under $7 a bottle, futsu-shu

33

PREMIUM-GRADE SAKE

There are essentially eight broad categories of sake deemed "premium." Some are easy to tell apart, while others require a sharper nose and palate to distinguish from other premium grades. For the purposes of this book, "premium" sake is defined as any product categorized within one of the following: honjozo, tokubetsu honjozo, tokubetsu junmai, ginjo, daiginjo, junmai ginjo, and junmai daiginjo.

→ HONJOZO: The most misunderstood style category within sake, in my gaijin opinion, has to be honjozo. And that's the fault of one of its defining characteristics: the addition of a small amount of distilled alcohol to the sake mash. While some producers today may still add spirit to reduce costs as was first done hundreds of years ago, many fortify their sake with distilled alcohol for aesthetic purposes. The added alcohol helps concentrate the aroma and gives the beverage an oily, silky mouthfeel.

For the beverage to be classified as honjozo, its seimaibuai must be no more than 70 percent (at least 30 percent of the rice kernel polished away). Additionally, the amount of distilled alcohol added must not exceed 10 percent of the total weight of rice used in the mash. Anything more than that and the resulting brew is just a futsu-shu.

Now let's get something out of the way right now: adding spirit does not raise the ABV of the sake. A major source of frustration for many sommeliers and brand reps occurs when they're describing honjozo at a tasting and the taster says something like, "Oh yeah, I can feel the burn from the extra alcohol." That's just the power of suggestion, because the honjozo isn't any more alcoholic than, say, a junmai, junmai ginjo, or junmai daiginjo. The only slight exception is if the sake in question is a genshu—that is, it's bottled at its original fermentation strength of 18 or 19 percent ABV and not diluted down to 15 percent. A little more boozy heft may be detectable, but a difference of three or four percentage points is still pretty negligible.

The flavor and aroma of honjozo-grade sake can vary wildly.

→ TOKUBETSU HONJOZO: *Tokubetsu* is a vague term if there ever was one. A number of nebulous factors can lead to a honjozo being dubbed "tokubetsu," which is Japanese for "special." Sometimes it's the rice polishing ratio—where honjozo proper has a seimaibuai no greater than 70 percent, tokubetsu honjozo may be around 60 percent. In other cases, it could be the use of a special yeast strain that makes it so "tokubetsu."

⊪→ GINJO: Okay, wait a minute, you say—doesn't a seimaibuai of 60 percent and under make it ginjo grade? Well, if there's that "special brewing process," it's more meaningful to get that "tokubetsu" on the label for some producers. Also, ginjo has some style-specific production methods to achieve its signature light and fragrant flavor, usually involving particular yeast strains and lower brewing temperatures. Ginjo typically has a more floral and fruity aroma than the honjozo. If the label reads "ginjo" and not "junmai ginjo," it means there's some distilled alcohol added (again, no more than 10 percent of the total weight of the rice). And at the risk of sounding like a broken record, no, that does not make ginjo inferior to junmai ginjo.

⊪→ DAIGINJO: With a seimaibuai of 50 percent and under (50 percent or more polished away), daiginjo-grade sake typically has a more delicate flavor and an even more complex, fruity aroma than ginjo-grade nihonshu. Some brewers become seimaibuai daredevils and mill away so much of the rice that they end up with a daiginjo whose seimaibuai is down in the single digits (though it is exceedingly rare; expect a seimaibuai around 40 percent). And the usual rules apply here: if it doesn't say "junmai daiginjo," that means there's some spirit added (say it with me, "no more than 10 percent of the weight of the rice"). Brewers often craft daiginjo sake for competition. The added alcohol gives these highly polished sake more aroma and weight. "They are usually brewed as show-stoppers in a glass to stand out at a competition," says New York-based sake sommelier Jamie Graves, "as opposed to something that's meant to accompany a meal."

⊪→ JUNMAI : The term *junmai* does double duty as both a descriptor and a style unto itself. It means "pure rice" and applies to those premium-grade sakes without added distilled alcohol. When the word is not followed by "ginjo" or "daiginjo," it's typically a category for sakes boasting a much more pronounced flavor (versus aroma) than ginjo or daiginjo, but generally a higher seimaibuai. There's no exact standard for what its rice-polish ratio should be, but it tends to be no higher than 70 percent (often considered the threshold for the "premium" designation).

35

The flavor distinction can be striking. Umami frequently is the first word that comes to mind, but it's highly subjective—I've had some that have reminded me of plates of stir fried noodles (highly subjective; the same nihonshu likely evokes different dishes among other tasters).

Some junmai sakes benefit a great deal from warming, but that doesn't mean they're any less enjoyable at room temperature or slightly chilled. That's why I sometimes like to use the noodle descriptor—I enjoy eating cold Asian noodles as much as I do hot ones.

»→ TOKUBETSU JUNMAI: Many of the same elements that make tokubetsu honjozo so "special" do the same for tokubetsu junmai, sans the added alcohol. For starters, the rice is usually polished a bit more than the 70-ish percent common for premium junmai sake. Sometimes it can be well below 60 percent. From a flavor/aroma/texture standpoint, tokubetsu junmai is a bit smoother and more delicate than junmai proper and can be virtually identical to a junmai ginjo, thanks to their similar seimaibuai. But, as is the case with its honjozo counterpart, brewers often opt to classify it as tokubetsu junmai to set it apart from ginjo-grade offerings they already have in their portfolios.

»→ JUNMAI GINJO AND JUNMAI DAIGINJO: Now that you know what junmai, ginjo, and daiginjo are, it's pretty easy to extrapolate the definitions of junmai ginjo and junmai daiginjo. Both of those are pure rice sakes with no added distilled spirit. The same seimaibuai rules for ginjo and daiginjo proper apply here. It's not always simple to divine the difference, flavor- and aroma-wise, between ginjo- and daiginjo-grade sakes and their junmai equivalents. Slight textural difference may be detectable, as the addition of alcohol typically produces a silky, more oily mouthfeel.

Sticker shock is also a common factor with junmai daiginjo, especially in bars and restaurants. But again, don't fall victim to the "pricier is better" mind-set. It's more about the cost of the raw materials as well as the labor intensity of getting junmai daiginjo just right. It's a style that needs a lot more coddling than others because there's just so much that can go wrong. In order for the toji —the master brewer— to better control the brewing process, it's often produced in smaller vessels than ginjo or junmai ginjo.

Now we're going to venture beyond premium grades and into classifications that can exist across those grades.

→ NIGORI: Nigori sake isn't necessarily one style in and of itself, as it tends to straddle a number of grades. There can be a junmai ginjo nigori, junmai daiginjo nigori, honjozo nigori, futsu-shu nigori—you get the idea. Each of those adheres to its respective seimaibuai guidelines. The unifying element is that they're coarsely filtered, leaving tiny, solid rice particles in the finished product. Nigori is cloudy white, with a creamy mouthfeel and typically a sweeter taste (though I've had some nigoris that were medium-dry to full-on dry).

The only real advice I have for drinking a nigori is that if you're pouring it yourself, shake the bottle beforehand to ensure that the little rice bits disperse and don't rest on the bottom of the bottle.

And remember to always avoid using the term *unfiltered* to describe nigori sake.

→ NAMAZAKE, NAMACHOZO, AND NAMAZUME: The advent of pasteurization was a real game-changer for every fermented beverage category, and that's no less true for sake. But that doesn't mean there aren't some highly desirable forms of sake that eschew the high-temperature, bacteria-eradicating process.

37

The majority of commercial sakes go through two levels of pasteurization—or hi-ire. The first occurs immediately after the filtration process and the second, immediately before bottling. The industry classifies any product that skips first, second, or both hi-ire stages as draft sake. And there are three designations under the broader draft umbrella: namazake, namachozo, and namazume.

Namazake is completely unpasteurized; it bypasses both hi-ire stages. When brewers want their finished product to be namachozo, they abstain from the post-filtration hi-ire but pasteurize just before it goes in the bottle. The reverse is true for namazume—it only skips the bottling stage hi-ire.

Sound arbitrary? As you'll soon find upon completing this book, there's nothing arbitrary about sake making. There's always a method to the seeming madness.

Namazake—or nama-nama, as it is frequently called—is likely
to taste and smell the most "fresh." Another way to describe it is
"bright."

Expect namazume's flavor to be a bit mellower than namazake, but
with some of those popping, fresh elements still intact. It's usually
aged for six months after the first and only hi-ire. The absence of
the second hi-ire allows the nama to retain some of the namazake
character.

Meanwhile, you might detect more umami notes in namachozo
than in namazake.

Flavors within all three, of course, will vary greatly, depending on
other, non-pasteurization-related processes. Rice polishing ratio
plays a role, as there are unpasteurized versions of any of the aforementioned grades.

You may have to ask the retail clerk, the sommelier, or the bar
manager which specific type of nama you're ordering. More often
than not, a product is labeled simply as "draft," "nama," or "nama
sake," not specifying whether it's namazake (or nama nama),
namachozo, or namazume. That's not a nefarious attempt at obfus-
cation, though; it's merely about label simplicity.

Since nama sakes are essentially alive, it's critical that they be
stored in cool temperatures and served soon after purchase. If
there's anything left in a bottle of namazake after a drinking ses-
sion, you really want to refrigerate it and finish it within the next
three or four days.

True namazake produced in Japan have very limited availability in
the United States. Oftentimes specialty retailers and more upscale
Japanese eateries will sell imported nama nama during a narrow
time window—usually in spring, at the conclusion of the traditional
brewing season.

"Nama season is always very exciting for us," says Kate Koo,
co-owner of Zilla, Portland, Oregon's premier sake and sushi
venue. "They're always limited runs and we get them allocated to
us. We really like to push those through during nama season so
people can really experience the style. A lot of times we'll have the
pasteurized versions as well so they can compare."

But you don't necessarily have to wait for nama season. Improvements in packaging, shipping and storage have made it possible for some nama to be available year-round. Just be wary of any restaurant or retailer that's storing it in warm places (I've had some that were rendered undrinkable for that very reason).

SUBCLASSIFICATIONS

Within many of the aforementioned styles and grades, there are plenty of sub-classifications, depending on the processes the brewers employ to produce them. For instance, you might see a label that reads "junmai ginjo yamahai" or "tokubetsu junmai muroka." Yamahai involves manual propagation of lactic acid–producing bacteria (versus adding ready-made lactic acid) to achieve a richer flavor profile, while "muroka" means it foregoes filtration.

SSI FLAVOR CATEGORIES

The Sake Service Institute (SSI), a leading Japanese organization for sake education and sommelier certification, has created a system to describe flavor and aroma characteristics of all of the grades and styles we just discussed. To that end, SSI developed four terms to help servers articulate different sake properties: kun-shu, so-shu, jun-shu and juku-shu. A certain ginjo brand, for instance, might exhibit flavor and aroma characteristics of a so-shu, while another might possess attributes that put it firmly within kun-shu territory.

Keep in mind that the four tasting terms are very specific to SSI's curriculum and are not standardized across the industry. In other words, only a relatively small group of servers, sommeliers and retailers use them. In fact, if you were to walk into a shop and ask for a kun-shu, so-shu, jun-shu or a juku-shu, it's very likely that the manager's response will be a blank stare if that manager has not gone through SSI-guided training. And you won't find those terms on a bottle's label. But I feel they are worth mentioning in this section because they're very helpful in describing the hard-to-describe and are terms that were part of my own sake education. (So maybe I'm a bit biased.)

Keep in mind that tasting is very subjective and no answer can be completely right or wrong. Still, some answers can miss the mark more than others. Let's explore the wonderful, if esoteric, worlds of kun-shu, so-shu, jun-shu, and juku-shu.

»→ KUN-SHU: Simply put, kun-shu is all about aroma. I'd even go so far as to say that flavor takes a back seat to what's on the nose. Typically delicate on the palate, a kun-shu is best enjoyed on its own to avoid it being overwhelmed by whatever it is you're eating.

Intense floral and fruity aromas are typical of kun-shu sakes. That might sound familiar—those are the notes common in daiginjo- and ginjo-grade (junmai and non-junmai versions) nihonshu—and they're usually the only grades that earn the kun-shu descriptor. But they don't fit into the category equally. Most daiginjo and jun-mai daiginjo sakes are kun-shu, while many ginjo and junmai ginjo bear that distinction. Many within those two latter style classifica-tions can also be right at home in the next flavor category.

»→ SO-SHU: The term *so-shu* is reserved for sakes that are more about refreshment than aroma. Many of those fruity and floral ele-ments are still present; they're just not nearly as pronounced. The best way to describe a so-shu is "light in flavor, light in aroma." They can be quite smooth, but they can also hold a distinct dryness that might create a sharp feeling on the palate.

So-shus are also far more flexible in the food-pairing arena. They tend to complement lighter fare like white fish, vegetable tempura, salads, and dishes commonly consumed as appetizers.

While so-shu is usually ginjo's turf, some daiginjo wouldn't be out of place here. Tokubetsu honjozo also has a home here, especially since seimaibuai-wise, the style is virtually identical to ginjo. The only difference, of course, is the addition of distilled alcohol to the former.

»→ JUN-SHU: The kun-shu quadrant might have the most elegant and delicate sakes, but jun-shu, at least in my opinion, has the most fun. A jun-shu sake usually has a subtle aroma of cooked rice, nuts, fried noodles, and sometimes little hints of funky cheese and yogurt. The texture is quite rich with big hits of umami on the palate. And, as you might surmise, jun-shu can be a sommelier's best friend when it comes to finding appropriate pairings for an array of fuller-flavored main courses. It matches well with many of the same items with which you might drink a full-bodied Cabernet Sauvignon—red meat, stew, and, naturally, a wide swath of tradi-tional Japanese fare.

I've even found a jun-shu to be a terrific match for New Jersey's greatest culinary contribution: the illogically named (and spelled) Texas Weiner, a deep-fried hot dog topped with mustard and chopped raw onions and smothered in bean-less chili sauce. I lived most of my life in the Garden State, so maybe I'm just biased.

→ JUKU-SHU: Juku-shu is the narrowest of the flavor categories. The designation applies mainly to aged sakes. Its flavors and aroma are robust and complex, with many of the same elements of jun-shu, but with more caramel and spicy notes imparted during the maturation period. Where the other styles (save for their nigori versions) usually range from transparent and colorless to faintly golden or yellow, juku-shu can be full-on amber.

TO EVERYTHING, THERE IS A SEASON

For culinary guidance, all you really need to remember is that aged goes with aged: aged cheeses, meats, etc. A juku-shu also holds up quite well against wild game.

Most sakes taste good no matter what season it is. But if you're keen to build a menu around a particular time of year, you can't go wrong if you follow these simple hints.

→ SPRING: Kun-shu and spring have a natural kinship. There's a freshness to a ginjo or daiginjo that exudes "springtime." It's pretty intuitive when you consider that one of the principal aromas is a bouquet of flowers.

→ SUMMER: We crave refreshment in the summer, and there's no flavor style more refreshing than so-shu.

→ FALL: The air gets crisper, and we move away from the salads and grilled meats toward richer, heartier fare like roasts. You might want to start thinking about a jun-shu at this point.

A QUICK CAVEAT

→ WINTER: Jun-shu's equally at home here as it holds up nicely with meaty stews that warm us up in those frigid months. The season also belongs to juku-shu, as its aged complexity makes sakes of this sort an ideal partner for wild game and the like.

One thing to keep in mind when taking some of this terminology out for a spin is that kun-shu, jun-shu, so-shu, and juku-shu aren't exactly discrete categories. The industry usually expresses them as a matrix, a graph with intersecting x and y axes, divided into four equal quadrants. A sake can fall anywhere in those quadrants and even straddle a couple different quadrants. When I took my kikisake-shi class, one of our tasks was to blind taste and determine the appropriate flavor quadrant. If one of us said "so-shu," the instructor would take it a step further and say something like, "I'd go with 70 percent so-shu and 30 percent jun-shu." It was actually quite maddening, but it illustrated that this is all a continuum and nothing is ever really 100 percent one thing or another. You know, kind of like people, and life in general.

CHAPTER 5

TASTING NOTES: THE NAMA-SPHERE AND GENSHU-VERSE

asteurization and dilution be damned! This next batch of samples is all about the various iterations of nama, genshu, and in some cases, a combination of the two!

MATSUNO KOTOBUKI NAMACHOZO

Brewer: Matsui Shuzoten
Prefecture: Tochigi
SMV: +6

This namachozo (the sort that bypasses the first of two rounds of pasteurization) isn't big on surprises, but I mean that as a compliment. It acts like many a fine ginjo-grade sake with the requisite fruit notes (in this case, honeydew), tempered by a little licorice. Its oily consistency makes it a bit more memorable than others of its ilk.

KUR-DASHI NAMAGENSHU

Brewer: Yamamoto-Honke
Prefecture: Kyoto

The Fushimi izakaya/yakitori pub Torisei pours this unfiltered, unpasteurized sake from a large tank in its main drinking and dining room. It's sharp and dry with fruit notes that lie at the nexus of melon and pear, and it goes great with just about any of Torisei's meat and veggies on sticks.

HINODEZAKARI JUNMAI GINJO MOMO NO SHIZUKU

Brewer: Matsumoto Shuzo
Prefecture: Kyoto

Only available from December through March, Momo no Shizuku ("Dewdrops of Peach") is the brewery's limited edition nama that attempts to capture the freshness of winter. Its sharpness is definitely reminiscent of the season's biting cold.

KIZAKURA GINJO NAMAZAKE

Brewer: Kizakura Shuzo
Prefecture: Kyoto
SMV: +2

Hope you like bananas, because this one is bursting with an aroma that's unmistakably reminiscent of the fruit. That doesn't mean it's sweet, however. It's quite dry—I'd even say drier than its SMV of +2 would have you believe—but it also manages to be soft and round at the same time.

TAMANOHIKARI JUNMAI GINJO SHIBORITATE

Brewer: Tamanohikari Shuzo
Prefecture: Kyoto
SMV: +1

The acidity gives this nama genshu a perceived dryness that's far greater than the official SMV of +1. It also accentuates some notes of sour apple that come through on the nose as well as a hint of cherry blossom.

KAGIYA NAMAZUME

Brewer: Seiryo Shuzo
Prefecture: Hiroshima

There's a stark minerality and a little citrus character in this nama-zume, but mostly the junmai is a sharp and rocky affair.

FUKUCHO "MOON ON THE WATER"

Brewer: Imada Shuzo
Prefecture: Hiroshima
SMV: +3

Melon, pineapple, and citrus dance with hints of black licorice in a silky junmai ginjo that made me think, for a very fleeting moment, that I was drinking the liquid equivalent of a Good & Plenty candy. And that's no dig. This was quite a memorable imbibing moment.

45

HIDEYOSHI NAMACHO HONJOZO

Brewer: Suzuki Shuzoten
Prefecture: Akita
SMV: +0.5

This semi-dry to mildly sweet sake tilts toward nutty, ricey graininess, but there are some subtle hints of tropical fruit if you spend enough time nosing it. It's not likely to overpower a salad, so it's great during an appetizer course. It also makes a good partner for grilled chicken.

BORN MUROKA NAMA GENSHU

Brewer: Katoukichibee Shouten
Prefecture: Fukui
SMV: +4

Sour pineapple, banana, and pear mingle on the nose of this genshu (undiluted—ABV is between 17 and 18 percent). There's a little bit of blue cheese expressing itself, likely due to the lack of pasteurization. And there's something comforting about its smooth, velvety texture that blankets the tongue.

NAMAZAKE NO JOZEN MIZUNO GOTOSHI

Brewer: Shirataki Shuzo
Prefecture: Niigata
SMV: +5

A namazake (completely unpasteurized) from Niigata? Yes, please!

A little bit of rice and a fleeting umami character offset the dominant tropical fruit notes to an extent. There's a succinct sweetness up front, but the balance of this journey is clean and dry.

KIKUSUI FUNAGUCHI

Brewer: Kikusui Shuzo
Prefecture: Niigata
SMV: -2

The least interesting thing about this nama genshu is the bright, golden can that holds it (and it's a pretty damned cool package!). Notes of steamed rice and banana bread are the star players in this non-charcoal-filtered honjozo. It's smooth up front with a sharp spike that kind of ambushes the palate.

NIIGATAMEIJO JUNMAI GINJO

Brewer: Koshi no Kanchubai
Prefecture: Niigata Meijo Co. Ltd.
SMV: +4

At 18 percent ABV, it's a genshu—a sharply acidic and citrusy one at that. But the element that will stay with you is the fact that it, well, stays with you. In other words, the finish is extremely long. You can almost hear it shouting, "I'm not done yet!"

HAKKAISAN YUKIMURO

Brewer: Hakkaisan Shuzo
Prefecture: Niigata
SMV: -1

The real story with this junmai ginjo genshu is its aging method. It rests for three years in an insulated room (the namesake "yukimuro") packed with snow at a naturally consistent temperature of 3 degrees Celsius (about 37 degrees Fahrenheit). The snow-aging process results in a soft and smooth sake with a flavor so delicate, you'd almost forget the 17 percent ABV. The nose tilts toward fruity with a somewhat strong sweetness on the palate up front, but with a medium-dry finish.

MANOTSURU "FOUR DIAMONDS"

Brewer: Obata Shuzo
Prefecture: Niigata
SMV: +3

I don't think I'm ever going to be able to top online retailer Sake Social's description for this rather expressive junmai ginjo genshu: "One word to describe this sake brew: Feisty." (It made me realize I really need to up my adjective game when writing these tasting blurbs). I'll see your "feisty" and raise it an "exuberant." The nose can be intensely fruity at times, with tons of uber-ripe melon and maybe a bit of orchard fruit. Then, around the second or third sniff, I started to get those rice notes—not necessarily the sweet and steamed sort, but the toasted or fried and caramelized sort. Sometimes it evokes biting into a caramel apple. But even with all of that going on, the texture is incredibly smooth.

47

HARADA MUROKA JUNMAI GINJO

Brewer: Hatsumomiji Shuzo
Prefecture: Yamaguchi
SMV: -2

This junmai ginjo has quite a few asterisks next to it: it's a genshu, a muroka (non-charcoal-filtered), and unpasteurized. Its nose is floral and fruity—maybe a touch of honeysuckle and citrus—which accentuates its sweetness. The acidity, however, gives it a dry, biting finish.

KOSHI NO ISO JUNMAI GINJO MUROKA GENSHU

Brewer: Koshi no Iso Shuzo
Prefecture: Fukui
SMV: +2

The name might look long, but it kind of has to be because there's just so much going on in this robust sake. It has a lot of the familiar fruity, pineapple, and melon stuff happening, but there are also a few hints of licorice and banana—not actual banana, but more of that confectionery, popsicle sort. The latter adds a soft sweetness.

THE PRODUCTION PROCESS

I don't want to go too far into the technical weeds with the sake production process, but there are a few things that every nihonshu appreciator—from newbie to seasoned pro—should know. I promised I would talk a little more about sandan-jikomi (the three-step brewing process), and I'm going to do so in the least eye-glazing way possible. I'm not expecting anyone to go out and start a commercial sake brewery with the information gleaned from this book, but you will seem like the smartest person on the tour if you're familiar with this stuff.

Assume all of the necessary rice has been harvested. Fresh-picked rice kernels aren't that pristine, white variety you get at a restaurant when you forget to ask for brown rice. In its rawest state, rice is brown, and the outer layers need to be milled away to get to the starch at the center. All of that outer matter is mostly proteins, lipids, and minerals. You don't want too much of that stuff getting in your sake because it can make it rather unpleasant.

Before the invention of modern rice polishing machines (known as seimaiki) in the 1930s, it was virtually impossible to polish the rice to the ginjo or daiginjo level without breaking it. Those grades wouldn't exist had it not been for the seimaiki revolution.

Rice is fed into the vertical polishing machines and then passed through two rapidly spinning grindstones known as Kongo rolls. Once the kernels fall to the bottom, they're returned back to the top via conveyor belt to repeat the milling process (call it a Kongo line—or not). It often takes days before the rice reaches the desired seimaibuai. If the rice is being milled to a daiginjo level, it usually takes around 50 to 55 hours to get to the requisite 50 percent seimaibuai. When the brewers aim for a polish ratio of 35 percent, they typically have to monitor the process for about 80 hours.

49

You might be surprised to learn (as I was) that the vast majority of sake breweries in Japan—possibly 90 percent or more—don't do their own polishing. They outsource to companies created specifically for that purpose. A shuzo without a polishing machine would dictate the specifications—type of rice, seimaibuai, etc.—and the milling company would deliver the desired grain.

Those shuzos that do possess their own polishing equipment take immense pride in that fact. Kiminoi Shuzo in the town of Myoko in the Niigata Prefecture has a trio of mechanical polishers that practically get their own large room. The toji told me that it's very important for them to control the entire process, as that's the only way the company can be sure they're getting exactly what they want (and that they don't receive any broken grains, which can severely impact flavor).

Naturally, the portion of the rice that's ground off doesn't just disappear. The ample powdery residue that remains—known as nuka—gets a new life as either cattle and pig feed (the brownish-amber outer layers of the rice) or in various forms of human food (the finer, whiter inner layers). Niigata is quite famous for its rice crackers, which are made from white nuka. The powder is also frequently used in pickling.

Once the polishing process is completed and a period of cooling has ended, the rice is washed to remove any lingering particles still attached to the grain. The washing process is called senmai. Most breweries today use machines to wash their rice, but for some of their products—particularly those involving grain with a very low seimaibuai—washing by hand is preferred. The more you mill a grain, the more fragile it gets, especially when you get into the daiginjo zone. Washing by hand helps mitigate that, even though machine-washing is likely to get the grain cleaner.

And that can be as painstaking a process as it sounds. I witnessed manual senmai at Aoki Shuzo in the town of Uonuma. The washer filled a large colander-like vessel with polished rice and then immersed it in a large blue bucket filled about a foot and a half deep with water. He'd let it sit there for a few seconds and pull it out, letting the spent water drain back into the pail and then repeat that procedure several times, using several buckets full of clean water.

Washing (senmai) is not to be confused with soaking (shinseki). The rice soaks to absorb enough water for the steaming process, which is the next step. The brewers steam the grain for a period of about 40 to 60 minutes. They then cool it and divide it into two distinct groups:

the kojimai and the kakemai. As you've probably guessed, kojimai is the rice used to make the rice koji, and it accounts for about 20 percent of the steamed rice. The other 80 percent is the kakemai, the rice used for the moromi (the mash).

Before we get to fermentation of the moromi, let's explore the complex, seven-step, 48-hour process required to make the komekoji.

→→ HIKIKOMI: The first step in the rice-koji-mking process involves bringing the steamed rice to the kojimuro, which, you'll recall, is a room with a temperature strictly set to 35 degrees Celsius (95 degrees Fahrenheit). The rice is spread out on long tables and turned and mixed periodically—every seven to fifteen minutes or so—by hand. The turning and mixing procedure gradually lowers the rice's internal moisture content and temperature. The koji makers stick a thermometer probe into the rice to gauge the changing temperature.

→→ TOKOMOMI: Once they've achieved the optimal temperature, the workers sprinkle the koji mold—a process known as Tanekiri. They then roll the rice around a bit to make sure the fungus is distributed evenly. Once that's done, they wrap up the rice again to keep the moisture in and regulate the temperature.

→→ KIRIKAESHI: During the 12 or so hours after tokomomi, the rice starts to harden and clump together. Kirikaeshi is the process of loosening those lumps and then wrapping the rice again (some producers use a turbine-like machine that aids in the lump loosening).

→→ MORI: Another 10 to 12 hours after kirikaeshi, the rice is divided into smaller clusters and transferred to small boxes to better control the temperature. Many breweries have both stainless steel and cedar boxes, depending on the sake grade they hope to achieve. For the premium styles, especially daiginjo, they'll employ the wooden boxes.

→→ NAKASHIGOTO: About eight or nine hours later, members of the komekoji-making team toss and spread out the rice to even out the temperature (the kojikin generate some heat).

→→ HIMAI SHIGOTO: About seven hours later, the rice is tossed once again, but this time the brewery team forms little ditch-like grooves that allow excess moisture to evaporate.

51

»→ DEKOJI: Once enough koji has been cultivated, it's moved out of the kojimuro into a more temperate room to halt the process.

At most of the breweries I've visited, the tojis were eager to have me taste the rice koji—which always surprised me since it is such a careful, sterile process, and some didn't let any outsiders inside the kojimuro. Even more surprising was the fact that fully cultivated rice koji tasted quite sweet. Trust me, the room itself smells anything but sweet (after all, these are fungi we're dealing with). But I could have actually mistaken the fully funked-up rice for a dessert. It was quite the pleasant confection.

Now, a bit on the fermentation starter, known as shubo (also sometimes referred to as moto). The shubo needs to have enough lactic acid in it to kill off any unwanted bacteria that would have adverse effects on the sake. There are a couple of methods to create that environment in the shubo, the most common today being sokujo-kei shubo. Simply put, a brewer buys commercially available lactic acid and adds it to the rice, giving the yeast adequate cover as it goes forth and multiplies. The whole process takes about two weeks. About 90 percent of sakes produced today are done so through the sokujo-kei shubo. Up until 1910, though, all shubo making was of the sort known as kimoto-kei-shubo, where naturally occurring lactobacillus produce lactic acid of their own accord. That process usually takes about twice as long as the sokujo-kei method.

There are two distinct processes within kimoto-kei-shubo: kimoto jikomi and yamahai jikomi.

Let's tackle kimoto-kei-shubo first. Back in the day—let's call it the 1900s and earlier—brewery workers would poke the vats of rice with a long stick called a yamaoroshi in order to crush the rice kernels and release the starches needed to promote saccharification. Pretty tedious.

Fortunately, in 1909, Japan's National Institute of Brewing Research realized that all of that jabbing was unnecessary and that the use of a yamaoroshi could be eliminated. The Japanese phrase for "elimination of yamaoroshi" is "yamaoroshi haishi," which ultimately was compressed to "yamahai." The yamahai method developed around the time rice milling became much more advanced. Yamahai only works with rice that's been milled to 70 percent or less. Such a seimaibuai was unheard of before the turn of the twentieth century.

52

Of course, only a year later, those same good folks at the National Institute of Brewing came up with the sokujo-kei-shubo process.

Now about 10 percent of the sake on the market is made via kimoto-kei-shubo, and it's not just yamahai. There's still some kimoto jikomi happening as well (though there are nine times as many yamahai jikomi sakes as there are kimoto jikomi). Given the additional time and labor associated with kimoto-kei-shubo versus sokujo-kei-shubo, the former is considerably more expensive (even more so when it's kimoto jikomi).

It's a rather labor-intensive process as well that demands constant attention. The shubo temperature must be regulated to ensure the most hospitable environment for the lactic acid–making microbes. And that's accomplished, pretty much, by hand. The brewers use a heavy-ish cylinder known as a dakidaru to adjust the level of heat to the desired degree. They'll fill the dakidaru—made of either metal or wood—with hot water. Then they'll take it by its handle and partially submerge the base of the dakidaru into the shubo mixture, constantly turning it—alternating between clockwise and counterclockwise—to even out the temperature (not the easiest task, considering the dakidaru weighs about 20 pounds). The lactobacilli tend to work their best when the surface temperature is about 42 degrees Celsius. Other unwelcome microorganisms—not so much.

53

Few, if any, sake breweries are yamahai-exclusive. The ones that do have a yamahai program typically produce a considerably higher volume of sokujo-kei-shubo-based beverages. And the two systems usually don't even share the same room, as they require different ambient temperatures. When I visited Kiminoi, I immediately noticed the difference when I went from the yamahai room to the sokujo room. The former must be kept at around 4 degrees Celsius (40 degrees Fahrenheit), while the latter usually is 10 or so degrees warmer.

Needless to say, yamahai is an incredibly risky process with so many variables that must be carefully controlled—especially in the most critical first 14 days of the cycle. And even though the yamahai method takes twice as long as the sokujo procedure, the end product usually doesn't cost the consumer twice as much. There's some markup, obviously, but it's often not proportional to the brewer's extra cost and effort.

At this point, you're likely wondering why anyone is bothering with kimoto-kei-shubo, when a faster, cheaper, and much more manageable method was developed to replace it more than a century ago. As is the case with almost anything else nihonshu-related, it's all about flavor. If you're looking for a richer, deeper, sometimes funkier flavor, you might want to sip a kimoto-kei-shubo sake. Sokujo-kei-shubo's flavor and aroma tend to be relatively milder, broadly speaking.

Brewers and sommeliers recommend pairing yamahai with red meat and Chinese food, as its high level of acidity cuts through the fatty, oily elements that dominate such dishes.

> It's time to get into the heart
> of sandan jikomi, the three-step
> brewing process.

Once the shubo is ready, it's dumped into a vat and ready for the kakemai and rice koji. When combined, those elements become the moromi ("mash"), of which the shubo is about 6 percent. Then comes day one of sandan jikomi, when 20 percent of the total steamed rice, water, and rice koji are added to the shubo. This first-day addition is known as hatsuzoe. On the second day, no further rice, water, or rice koji are added—the moromi just sits and allows the yeast to propagate. This day of rest is known as odori. On day three, the brewers add a portion of rice, rice koji, and water equivalent to about 30 percent of the moromi to the mix. This stage is called nakazoe. Finally, on day four, it's tomezoe time, as the remaining 44 percent of rice, rice koji, and water go into the moromi.

Next up is a process that's also unique to Japanese alcohol: multiple parallel fermentation. The reason for the "parallel" is that both saccharification—converting starch into sugar—and fermentation happen simultaneously in the same tank. So, while the koji are working hard to break down all of the rice's starches into fermentable glucose, the yeast is already getting to work on the sugar to turn it into alcohol.

This process is not to be confused with single-line multiple fermentation, which is the method with which beer is produced. In beer, barley germinates through the malting process—the grains are soaked in water and then dried through kilning—which produces the enzymes necessary for saccharification. After that step, brewers add the yeast to get the fermentation going.

The process for wine, on the other hand, is simply single fermentation. That's because the grapes already contain the sugar that's going to be converted to alcohol, requiring no intermediate step.

The process of multiple parallel fermentation is the reason sake is naturally the highest ABV fermented, non-distilled beverage there is, at 18 or 19 percent before it's diluted to about 15 percent. You might be thinking, But I've had beers that were well over 20 percent ABV. It's true, such extreme, boozy brews exist, but they don't get to that strength with the malt base alone. Usually, additional sugar sources are added after the fermentation process has run its course to wake up the yeast and get it to continue doing its thing. It's almost like using a defibrillator on someone who's flat-lining. Sake, though, gets to its elevated ABV without any additional help.

About three days into the multiple fermentation process, streaks of foam start to appear, confirming that fermentation has indeed begun. By about the sixth day post-tomezoe, what started out as light, bubbly foam has transformed to a firmer foam that vaguely resembles moon rocks. At this point, a rotating propeller descends from the top of the tank to break up the foam before it rises too high. It reaches peak height between the eighth and tenth day. By the twelfth day, the foam starts to noticeably die down. Within a few days after that, all that's left are bubbly beads of foam on the surface of the moromi. Then, around the twentieth or twenty-first day, the bubbles completely vanish.

55

After multiple parallel fermentation runs its course, the sake makers may send it straight to pressing or, if it's honjozo, they first will add a bit of distilled alcohol to the mash (a process called aru-tenshu).

Now, on to pressing, or "joso." At this point, the moromi still has a lot of solid rice material in it that needs to be separated from the liquid that ultimately will become sake. The solids, known as lees or "kasu," are removed via a variety of methods ranging from traditional to more technologically enhanced.

The traditional process, which is still very much in use today, involves filling cotton bags with the moromi and laying them in a big wooden box called a fune. Initially, before actual pressing begins, the gravity of the stacked bags of mash forces some of the liquid to pour out on its own. That prepress, free-flowing liquid is known as arabashiri (not that you need me to throw another term at you).

A lid descends into the fune, applying a great deal of pressure to the porous cotton sacks and squeezing out the liquid, which drains through a hole at the bottom of the box.

There's also a more time-intensive process called fukurozuri, in which the sake makers dangle the moromi-filled cotton bags from a rope and allow the magic of gravity to accomplish the joso task without any additional pressure. It then slowly drips into a large, 18-liter jug (there's that 1.8 multiple again) called a tobin—the process of collecting into such a vessel is called tobingakoi. Generally, nihonshu made through fukurozuri is bottled at a higher price point, as the process is so time consuming and results in a much lower sake yield than the fune method (there's only so much gravity can do without the added pressure).

As is the case with kimoto-kei-shubo, there are flavor profile–related objectives justifying the hard work here. And there is of course a much more modern joso method that enlists the services of a large, accordion-like machine that presses the sake automatically. A hose feeds the fermented moromi into the machine where rubber bags inflate and squeeze the liquid out through multiple mesh panels, leaving the kasu behind in large, flat sheets that resemble giant slices of cheese (you can even break a few pieces off and eat it).

If you're worrying that all of those solid remnants go to waste, fret not. The kasu usually gets a second life in the culinary world. It's used for pickling everything from vegetables to fish, and it frequently finds its way into marinades and soups. Chefs have also discovered that when spread on a raw steak, it accelerates the dry-aging process.

Once the sake is done being pressed, it rests in a tank for a week and a half or so as sediments settle to the bottom. It's then filtered for clarification and to remove undesirable color and flavor elements. And then comes the first round of pasteurization (hi-ire), unless it's namazake or namachozo. You'll recall that there typically are two rounds of pasteurization, once before it goes in the tank and once at bottling, and that namazake goes through neither, while namachozo is pasteurized only at the bottling stage. Namazume, on the other hand, does get pasteurized prior to tank storage, but not at bottling.

The term *chozo*, not to be confused with *joso* (pressing) or *jozo* (alcohol added to the mash), refers to storage in the tank. It sits in the tank for anywhere between two weeks and a year at a constant temperature of 15 degrees Celsius (59 degrees Fahrenheit, to those of us who are metrically challenged).

After the requisite resting period, the producers blend sake from different tanks together, similar to the way distillers combine whiskeys from different barrels to achieve the desired flavor and aroma profile.

Next comes warimizu (diluting with water), unless the finished product is meant to be genshu (the undiluted sort). Then it's time for that second pasteurization/hi-ire (unless, of course, it's namazake or namazume).

Finally, it's time for bottling. But if it's a sparkling sake, the production process isn't quite finished. As is the case in beer and sparkling wine, this can be achieved one of two ways: through forced carbonation or, the natural method, through secondary fermentation in the bottle.

SUGIDAMA

It's not always easy to spot a sake brewery, especially when it's a producer of modest size located in a small town (and not being able to read Japanese kanji certainly doesn't help). A good clue that nihonshu is happening inside is the presence of what looks like spherical shrubbery dangling from the eaves of the building, which itself typically looks residential or retail-related. That mysterious sphere is known as a sugidama, which translates to "ball of cedar." Back in the day, the producers would hang these arrangements of fresh cedar branches early in the sake-making season when they made their first batch and stored it for aging. The leaves would gradually lose their green color, and when they became full-on brown, that meant the sake was just about ready. There are, of course, much more modern means of communicating this, but brewers continue to hang them outside their facilities for tradition's sake.

CHAPTER 7.

TASTING NOTES: YAMAHAI AND KIMOTO

We're going to kick it old-school with these next few . . .

TENGUMAI YAMAHAI JUNMAI "DANCE OF THE DEMON"

Brewer: Shata Shuzo
Prefecture: Ishikawa
SMV: +4

Any day you get to sip a yamahai sake is a good day. And it's worth risking eternal damnation to dance with this demon! It's everything a yamahai junmai should be—all of the cooked rice and toasted almond notes you'd come to expect, with abundant earthiness, a little bit of melon, and some creamy, rustic, stinky-cheese-like funk to keep you interested. The terms *full-bodied* and *rich* only begin to describe it.

SHIOKAWA COWBOY YAMAHAI

Brewer: Shiokawa Shuzo
Prefecture: Niigata
SMV: +3.6

If you're feeling like a cowboy, Shiokawa's yamahai genshu (18.8 percent ABV) will take you for a wild ride. It's buttery, with a velvety texture and an assertive umami quality. I'd say it's a bit of an acid bomb as well.

KAGATOBI YAMAHAI JUNMAI SUPER DRY

Brewer: Fukumitsuya Shuzo
Prefecture: Izikawa
SMV: +12

Very striking acidity helps further accentuate the intense dryness (hence, "super"). It's bold, rich, and crisp, with some roasted nut and cooked rice quality and even a little vinegar-like sour grapes (the tasty kind, not the whiny kind). The acidity should help it hold up pretty well against some of the fattiest, oiliest meals.

KIMINOI JUNMAI GINJO YAMAHAI

Brewer: Kiminoi Shuzo
Prefecture: Niigata

The brewers at Kiminoi were a bit taken aback when I told them their junmai ginjo yamahai was vaguely reminiscent of strawberry frozen yogurt. But it was coming from a place of love and utmost respect. They also admitted they're a bit wary of the new way for kikisake-shi to describe sake with fruit and food descriptors, which made me question everything I learned in my kikisake-shi class!

KIMINOI JUNMAI DAIGINJO YAMAHAI

Brewer: Kiminoi Shuzo
Prefecture: Niigata

Kiminoi takes the polishing a bit further with its 40 percent seimai-buai junmai daiginjo yamahai. And the result is a sake that's remarkably cleaner than its ginjo counterpart. There's less of the strawberry yogurt thing going on, but there's some definite berry character hanging around. I also sensed a bit of pear and maybe a little melon as well—and that's the sort of fruit salad I'd be happy to have any day.

TAMANOHIKARI JUNMAI GINJO YAMAHAI CLASSIC

Brewer: Tamanohikari Shuzo
Prefecture: Kyoto
SMV: +1

It's a classic, all right, and a full-bodied one at that. It has some seriously dense umami and notes of multigrain toast and an aged semihard, nutty cheese.

GUNMA IZUMI YAMAHAI HONJOZO

Brewer: Shimaoka Shuzo
Prefecture: Gunma
SMV: +3

You've really got to get your nose in the glass to pull out the aroma, but when you do, there's a small drop of fresh citrus and a touch of sweet rice. Once you sip, you will notice some of that requisite yamahai creaminess, but it's definitely one of the more subtle selections within that genre. If there's such a thing as "medium-rich," this is probably it.

61

TAKASHIMIZU KIMOTO TOKUBETSU JUNMAI

Brewer: Akita Shurui Seizoh Co. Ltd.
Prefecture: Akita
SMV: +3

You really get the sense that the lactobacilli were working pretty hard in the Akita Prefecture, because the first thing you notice on the initial sip is a very pronounced yet refreshing tartness. The closest comparison I could immediately think of was umeboshi, those pickled plums that often show up in the breakfast buffets at Japanese hotels. There's also a strange salinity that sporadically expresses itself, practically making this the sake equivalent of a German gose-style beer (not really, but those tart and salty elements are very evocative). Some hard-cheese characteristics make me think of Parmesan on a rice cracker, perhaps on a board with some dry-roasted nuts. It's a kimoto; did you expect this description to be boring?

DEWATSURU KIMOTO JUNMAI

Brewer: Akita Seishu Co.
Prefecture: Akita
SMV: +3

If you see Dewatsuru Kimoto on a menu, the first thing you need to ask your server is the temperature at which it's poured. Hopefully it's not chilled. If it is, you have two choices: order something else, or better yet, let it sit for about a half hour before you drink it. Dewatsuru Kimoto does its best at room temperature, or even a few degrees warmer. Once it reaches that temperature, you're going to immediately nose aromas of toasted cereal, like a big bowl of Grape Nuts or the burned grains scraped off the bottom of a rice cooker. There's an almost meaty quality to it as well, with hints of miso and dry sherry. It could almost be a steak sauce.

KASUMI TSURU KIMOTO EXTRA DRY

Brewer: Kasumi Tsuru Shuzo
Prefecture: Hyogo
SMV: +5

A kimoto is always a special experience, but a kimoto from Hyogo? Sign me up immediately (you'll catch my drift about the famous prefecture in the next chapter)! It's deep, dry, and umami-forward with elements reminiscent of a seaweed salad with toasted sesame seeds sprinkled on top, perhaps with a side of hard, nutty cheese. At one point while drinking this, I swear I could smell the sea.

CHAPTER 8.

SAKE-OGRAPHY I:
THE BIG FOUR

When someone asks you for one or two of the world's great wine regions, it shouldn't take you too long to respond with "Napa," "Bordeaux," or "Tuscany." When they ask the same question about sake, it might take a bit more time for most people to formulate an answer beyond "Japan." But there are parts of Japan that are every bit as important to the sake community as those iconic grape-growing destinations are for oenophiles. In this chapter we're going to hit the four most iconic areas first.

NADA, KOBE, HYOGO PREFECTURE

Mention the city of Kobe and the first thing that comes to most people's minds is the style of beef that bears the town's name. The Tajima-gyu breed of cattle that are the source of this carnivore's delight must be born and bred in the Hyogo Prefecture in order for it to even have a chance to one day be classified as Kobe beef. Kobe is the capital of the prefecture, hence the name. Sure, some of those cows are from Kobe proper, but how many pastures do you think you'll actually find in an urban area where about 1.5 million people live?

You will, however, find plenty of operating sake breweries within the city's 213 square miles. The greatest concentration of nihonshu producers can be found within Kobe's Nada district, a 12-square-mile ward that's home to around 40 sake breweries (that's more than three per square mile), responsible for about a third of all nihonshu being produced in Japan. It's not unheard of for a person to walk to every single one of those operations—though I wouldn't advise attempting to hit them all in a day (and they're not all open to visitors, so you wouldn't be able to have a look inside anyway).

Technically, Nada itself isn't even a single district but a collection of five neighborhoods known as the Nada Gogo ("five villages of Nada").

Kobe gained prominence as a port city, which enabled other parts of Japan to enjoy the premium sakes that made Nada breweries famous. By the seventeenth century, Edo—now modern-day Tokyo—had already become quite the populous city, with around a million inhabitants calling it home. That made it the biggest market for the major sake-producing areas in the western part of Japan. Among those were Ikeda, Amagasaki, Sakai, Itami, Nishinomiya, and of course, Nada. But it wasn't until about the mid-nineteenth century that Nada became the dominant supplier, usurping Itami, which, at its peak in the early 1800s, had been producing more than 20 million liters (Itami production began to steadily decline shortly after that).

Part of the reason for Nada's surge was that its brewers became early pioneers in the rice milling process. There were six rivers flowing down from Mount Rokko that powered water wheels used in milling. Lacking such innovations, Itami producers used a much more laborious process of stomping on the grains. Speaking of water, Nada was the source of the highly revered miyamizu variety. You can get a pretty good education on the history of the district and its importance in the country's sake-making legacy at museums such as the Hakutsuru Sake Brewery Museum, located on the campus of the modern brewery. There's also the Sawanotsuru Sake Museum, as well as the Kikumasamune Sake Museum, complete with vintage wooden vessels and tools used more than a century ago. And of course, the museums also offer tastings. One of the best treats is Kobe Konan Muko no Sato, a shop that specializes in pickles and other delicacies made with sake lees from the neighborhood's producers. It's a great place to get your umami on.

NIIGATA PREFECTURE

The Niigata Prefecture is incredibly scenic, with its snowcapped mountains visible from most points throughout the region. The snow is fairly abundant—some areas get more than 30 feet each year—and it's a big reason why there are as many as 90 breweries producing quality sake across the prefecture. When the snow melts and seeps into the ground, it becomes a pristine source of water for brewers. It also helps create some of the purest paddies in the world for the varieties of sake rice that grow across Niigata. Driving through the prefecture, you'll likely notice teepee-like arrays of sticks surrounding trees.

Those are designed to fend off the snow and keep it from destroying the branches.

While the number of sake breweries in the Niigata Prefecture today is impressive, in the early 2000s there were about 20 more. Consolidation and increasing competition from other beverages—younger Japanese consumers are pretty fond of beer, for instance—has taken a few players off the market. Still, Niigata is home to the largest number of sake breweries out of all of Japan's 47 prefectures. Most of the Niigata breweries that are still operating don't seem to be going away any time soon. Many have remained in the founding families for anywhere between 100 and 400 years, and the new generations who manage the businesses today take tremendous pride in continuing their centuries-old legacies. It also helps that per capita sake consumption in Niigata is about twice the national average.

Over those centuries, a certain "Niigata style" of sake emerged, thanks particularly to that aforementioned water. The soft, pristine water yields sake that is clean, crisp, dry, and light. However, when I visited a number of breweries there and spoke with the owners and the toji, many prided themselves on producing sake that was a little outside the norm for the prefecture. Niigata nihonshu is not generally known for pronounced umami character, but many of the family breweries produced varieties that had plenty of that hard-to-pin-down savory flavor pivotal to Japanese cuisine.

The best place in the world to sample Niigata sake, in my opinion, is a shop and tasting hall known as Ponshukan, located in the main train station in Niigata City, as well as the Echigo-Yuzawa station in the Niigata Prefecture town of Yuzawa. Plan to catch that later train, because once you step inside, you're not going to want to leave any time soon. You pay ¥500 (just shy of $5, depending on the exchange rate) and get five tasting tokens, as well as a small janomeno-style cup (the one with two blue circles at the bottom that makes the cup look like a snake's eye) to use. There's a wall with about 100 different sakes to try, and you simply drop a token in the slot and let the chosen sake pour into your cup. And if you find something you really like—and it's hard not to—the shop next door is very likely to have a bottle of it in stock.

If you weren't already convinced that Niigata Prefecture is a little slice of sake heaven, it's also home to the renowned Niigata Prefectural Sake Research Institute. Since 1930, the institute has studied everything from rice cultivation to weather to water sourcing in order to

65

better understand how all of those elements impact the quality of Japan's prized beverage.

The organization also partnered with the Niigata Agricultural Institute to develop new strains of sake rice. Among those is a hybrid of the most popular strains, Yamada Nishiki and Gohyaku Mangoku, called Koshi Tanrei. A growing number of breweries in the prefecture are employing the hybrid variety to make some of their more premium products (ginjo and daiginjo grades, in particular).

Most sake-producing prefectures have their own toji guilds, and as you'd expect, Niigata's is one of the most prestigious in all of Japan. The Echigo Toji, as it's called, is rooted in a tradition dating back at least to the early sixteenth century. The Echigo Toji, known for their expertise, perseverance, and unflappable work ethic, had a great deal of influence over toji from other regions learning the sake production trade. Ultimately, the Echigo Toji developed a signature style that came to define what became Niigata sake, harnessing the best elements of its climate and terroir.

Echigo Toji are revered throughout the country, and brewers in many prefectures outside of Niigata often hire them for their well-regarded skill set.

Niigata brewers benefited from the gradual evolution of the Japanese diet in the post-war decades. Initially after World War II, food shortages led consumers toward sweeter, more robust and full-flavored sake along the lines of the sort coming out of Hyogo and Kyoto. Since a large portion of the Japanese population didn't have much to eat, they sought drinks that offered foodier flavor elements, often in lieu of meals. Niigata brewers started adopting some of those non-Niigata-style characteristics into their products.

However, by the mid-1960s, many brewers had incorporated some Western culinary traditions into their diets and were looking for more nuanced sakes that were flexible enough to complement a wide variety of foods beyond historically umami-forward Japanese dishes. The quintessential "Niigata style" fit the bill perfectly and helped cement the prefecture's place as a leader of the modern market.

In fact, the Niigata Sake Institute encouraged brewers to mill their grains more than most others in the country to ensure the most delicate and subtle sake possible. You're likely to find an average seimaibuai of 38 percent for Niigata-produced sake versus the standard 50 percent for that highly polished grade.

In 1997, the Niigata Sake Brewers Association went so far as to establish a set of standards for the prefecture's producers. As you'd expect, for a sake to bear the "Niigata Sake" label, it must be brewed within the prefecture's borders, and the rice and water must be sourced there as well. The brewers also must submit the product to the association's quality-control committee for approval.

A NIIGATA BREWERY TOUR

Personally, I have spent more time in Niigata breweries than those in any other prefecture—primarily because there are just so many of them! I'd like to spotlight some of those and take you on a very abbreviated tour through some of the snow-covered region's top producers.

↠ Kinshihai Shuzo

Among those was Kinshihai Shuzo, whose sake-brewing heritage spans two centuries in the north-central Niigata town of Gosen (formerly known as Muramatsu). Prior to their purchase of a brewing license in 1824, the founding Shigeno family—whose descendants still run the brewery today—worked as successful rice distributors. They decided to turn some of their excess inventory into sake, and when they realized it would be an even more lucrative endeavor, they switched to nihonshu-making full-time. It made both logistical and financial sense: Gosen ("five fountains") is situated at the edge of both a rice field and a mountain, with snow eventually becoming the pristine spring water—known as tengunomizu—used in Kinshihai's products.

The Shigeno family launched the company as Shigeno-i Shuzo but changed the name to Kinshihai in the mid-twentieth century when they were looking to refresh the family brand with a bold, new moniker. They settled on Kinshihai, which means "golden raptor"—can't get much bolder than that.

Kinshihai was an early pioneer in the burgeoning daiginjo market post–World War II. At the time, daiginjo barely existed as a style because it was incredibly costly and time consuming to produce—not to mention practically brand new since modern rice-milling machinery didn't arrive on the scene until the 1930s. Prior to that, it was virtually impossible to polish away half of the grain without breaking it.

67

In the 1950s, Kinshihai's president spearheaded a project to study and master daiginjo making; and since then, researchers and brewers from all over Japan have visited the brewery to enhance their own education.

The project team also developed its own bespoke yeast strain, which Kinshihai uses almost exclusively.

For six decades, the brewery has continued to produce at least one daiginjo each year to ensure that new and future generations never lack the expertise of their predecessors.

Kinshihai's products share the characteristics of Niigata sake, but the owners and the toji pride themselves on the distinctive umami richness typically detectable with a few whiffs and sips. General manager Takako Shigeno draws a parallel between her company's sake and the folks who inhabit the brewery's sleepy hometown: calm and reserved, not big and boisterous. Similarly, Kinshihai's offerings have a quiet balance to them: not too dry and not too full-bodied.

Shigeno also insists her company's sake plays very well with others and doesn't like to be alone. In other words, it's made to drink with a meal, not on its own. Kinshihai pairs with a multicourse menu from something as delicate as sashimi, to richer fare soaked in sweet and salty sauces. Gosen locals love it with the area's signature carp dishes, simmered in soy sauce, some sugar, spices, and herbs, which are consumed mostly on special occasions.

↠ Musashino Shuzo

Members of the founding Kobayashi family like to joke that Musashino is the baby of the sake brewing scene in Niigata Prefecture. Of course, that infant is more than a century old, as it first opened its doors in 1916. Now in their fourth generation, the Kobayashis remain in charge. I can't imagine why they'd ever want to leave; if they ever get too stressed, they can peek out the window at the beautifully manicured Japanese garden that surrounds their headquarters in the city of Joetsu. I asked managing director and sake master Hisashi Kobayashi if he employs a regular groundskeeper, and he responded, "Yes, me!" I do not know where he finds the time.

The other big star at Musashino is toji Kenji Fujii, who boasts one of the longest tenures throughout the prefecture. He's been brewing at Musashino for more than four decades.

Drinkers outside of Joetsu are most familiar with Musashino's Ten To Chi "Heaven and Earth" junmai daiginjo, made from Niigata's homegrown rice variety Koshi Tanrei. You'll often hear brewers and sommeliers recommend that you drink a junmai daiginjo on its own to fully appreciate all of its delicate nuances. "Heaven and Earth" does complement a variety of foods fairly well, however, thanks to some robust acidity that cuts through the fats of richer dishes. Other offerings of note include Daku nigori, which is sweet enough to pair with numerous dishes, and the fruity Into Your Soul honjozo, which is a good match for spicier fare.

Kobayashi escorted me a little off the beaten path when he broke out a 33-year-old mirin—the stuff that's usually sold as sweet cooking sake in the Asian food aisle. Usually, mirin sports a pale, yellowish hue, but I discovered what happens to that color after more than three decades of sequestration: it turns an opaque, extremely viscous, dark brown. It's still a little on the sweet side, but there are also elements that weren't unlike Worcestershire sauce, tobacco, and even chargrilled beef (note to self: I have to try this in a Bloody Mary sometime).

⇥ Matsunoi Shuzo

The quiet, snowy southwestern Niigata town of Tokomachi is the home base for Matsunoi Shuzo, founded in 1895. Like Musashino, Matsunoi Shuzo is now in its fourth generation of family ownership, under the leadership of CEO Makoto Furusawa. Matsunoi is definitely one of the more rural breweries I encountered in the prefecture. Twenty-first-century equipment provides a striking contrast inside a wooden-beamed, high-ceilinged structure that more closely resembles a pre-industrial barn than a modern production plant. In a sense, the building is a metaphor for much of the traditional sake business throughout Japan: a beast whose heart is in the past but whose mind is in the future.

Its sake-making processes are also a mix of old and new. A few feet from its modern Yabuta pressing machine sits a traditional fune apparatus—the pressing method that involves applying pressure to cotton bags of mash arranged across a wooden box to squeeze the liquid from the solid. For Matsunoi's daiginjo-grade products, it removes the pressure altogether, instead, employing the traditional fukurozori method (gravity separates the liquid from the solids in dangling cotton bags). Matsunoi is also in the minority of breweries with their own rice-milling equipment. The management prefers to polish its own grain rather than outsource it to a third-party milling company, as it enables Matsunoi to control nearly every step of the process and get the exact rice it wants.

The brewery's efforts have borne plenty of fruit; it's one of the most award-winning shuzos in all of Japan, having earned gold medals at the Japan National Sake Competition in all but three years between 2007 and 2015.

»→ Kiminoi Shuzo

Kiminoi Shuzo's story bridges two very important epochs in Japanese history: the Edo and Meiji periods. In 1842, during the waning years of the Edo period, the Tanaka family founded Kiminoi Shuzo in what is now Myoko City (formerly Arai) in the southern-central edge of Niigata Prefecture. During the Edo period, the city was a popular spot for traveling shogun to stop and rest, eat, and drink on their way to what is now Tokyo. The brewery's location on the main thoroughfare—where it remains today—made it the watering hole of choice on such journeys. It also was the primary route merchants used on their way to Edo to deliver everything from seafood to salt to gold.

But the company earned the name "Kiminoi" during the subsequent Meiji period, when the Japanese emperor would make the trek to Arai and pray at a temple right next door to the sake production facility. Naturally, the monarch took a liking to the beverage being produced adjacent to his place of worship, and the brewery soon became known as "Emperor's Well," or "Kiminoi."

Near Myoko is the mountain of the same name, which is responsible for the snow that ultimately becomes the precious source for Kiminoi. While snow is prevalent throughout most of the prefecture, residents of the Miyoko area would swear the white stuff was invented in their city. Brewery workers told me that the region has sometimes been known to get as much as 30 feet—yes, feet—on a single night. The massive, extremely dense wooden beams that support the brewery ceiling tell you everything you need to know about the weather.

Kiminoi, like Matsunoi, is very proud of its in-house rice milling operation, which is why the toji makes it the first stop on my tour of the production facility.

Kiminoi's polish-your-own mentality is really a pragmatic one, its managers assure me. When a brewery orders rice from a third-party milling operation, the only way to gauge whether its grain has been milled to the precise specifications is to weigh it. But there's really know way to verify the percentage. In other words, if a kernel of rice weighs 50 percent less than that of unaltered brown rice—"genmai"—that doesn't necessarily mean that 50 percent of it has been polished (my head spun too when I learned that detail). Also, if the quality of the genmai being milled is subpar, it's likely to break into pieces and be useless. And it's hard to find those broken bits inside gargantuan sacks of rice.

A difference of just 2 percent in the amount of protein and fats that remain on the rice can make a monumental difference in flavor and aroma. So, for Kiminoi, managing the milling really means ensuring a consistent product from batch to batch.

As for the rice itself, it's about as local as can be for a midsize nihonshu maker. Of the 120,000 kilograms (about 265,000 pounds) of grain Kiminoi goes through each year, half of it is from Miyoko and the other half is from elsewhere within the prefecture, mostly in the Joetsu area.

Another point of pride for Kiminoi Shuzo is its extensive yamahai program. The bulk of the brewery's yeast starter is made through the modern sokujo-kei-shubo method, but the size of Kiminoi's yamahai room—about half the size of the sokujo room—is still fairly large compared to others I've seen. The sokujo area is kept fairly chilly, around 65 degrees Fahrenheit, so I wasn't prepared for the huge, walk-in fridge that is the yamahai room, which is about 39 degrees Fahrenheit. And that temperature is kept constant. As much as I wanted to geek out in the yamahai room, I was in a bit of a hurry to get out because the only thing protecting me from the chill was a thin mesh disposable lab coat.

⟫→ Aoki Shuzo

For some brewers in the northern prefectures, the surrounding snow becomes a natural component of the company's production infrastructure. It's nature's own climate control—both refrigerator and cellar.

That's the situation at Aoki Shuzo in Uonuma City, the only brewery I've visited that let me play in the snow. Okay, not really play, but I did visit its snow room—*yukimuro* in Japanese—at the end of November, about a month and a half after it had been filled to its 400-ton capacity (the beauty of being located so close to Mount Makihata is that you can get your hands on some fresh powder in October). The metal walls quickly pick up the chill and help keep the mound from melting too quickly. Come January, the brewery will bring in another 400 tons.

Pumping the snow—or, rather, its liquid content—through pipes keeps the adjacent snow-aging room at a constant 0 degrees Celsius (32 degrees Fahrenheit, freezing point).

As you can guess, the yukimuro is quite an energy-efficient refrigeration method. And thanks to government tax credits for breweries that build them to dramatically lower their electricity usage, yukimuros have become quite common across Niigata Prefecture.

Aoki Shuzo is one of the older breweries in the region, having never stopped producing since opening doors in 1717. Today it is known for two signature brands, Kakurei and Yukiotoko. Aoki crafted Kakurei to pair with the local cuisine of Uonuma, whose cooks are known for their liberal use of salt and soy sauce (not only for flavor, but as a preservative for those long, snowy winters).

Yukiotoko has a further connection to wintry weather, as it's named for the mythical "snow yeti" that appeared in local eighteenth- and nineteenth-century author Bokushi Suzuki's book *Hokuetsu Seppu*. (Incidentally, the brewery says that Suzuki also may have been the one who gave Kakurei its name.)

Aoki Shuzo's philosophy can be summed up in one word: *wago*, the interconnectedness of everyone involved in sake. That's not just limited to the brewers but to the rice farmers, the merchants, and of course the drinkers. It's the "spirit of cooperation" and "spirit of endurance" that are defining characteristics of the inhabitants of Niigata Prefecture.

⇥ Midorikawa Shuzo
Founded in 1884, Midorikawa sits at the heart of rice country. In fact, its current location in Uonuma had been restricted to rice farming, but the brewery was able to get a special dispensation from the government to operate a sake-making facility there. Midorikawa president and Niigata Sake Brewers Association chairman Shunji Odaira notes that the locally grown variety is the most expensive grain in the country; the shuzo produces about half of its total volume from rice grown in the immediate area.

While a snow room is a progressive innovation for sake production, a snow dome is on another level entirely. Midorikawa (which means "Green River") ages some of its products in these igloo-like structures, which can help preserve and enhance the quality of the final product. As Odaira explains it, pasteurization can have an adverse effect on the flavors and aromas in the batch of sake. Aging in a snow dome for a couple months helps restore those elements to the sake. It plays an important role with unpasteurized sake as well. Since the quality of nama changes very quickly, providing it time in a snow dome slows that degradation and allows it to retain its optimal quality longer once it's bottled and shipped.

Midorikawa demonstrates that sake making is as much a blender's art as it is a brewer's. The company achieves its very precise flavor and aroma characteristics, particularly in its ginjo and daiginjo grades, by blending batches matured for a varying number of years. And Odaira admits that Midorikawa brands reflect his very singular tastes. He doesn't like nihonshu with big, noisy aromas, he tells me, because it doesn't pair well with American food. That's been a running theme among many producers in Japan. They're increasingly keen on getting away from sensory elements tailored only to the Japanese palate. Odaira describes his company's products as "calm" and "rich, but smooth."

➺ Kirinzan Shuzo
Three words come to mind when I think of Kirinzan Shuzo: *local, local, local.*

73

Of the rice the brewery uses in its products, 92 percent comes from its hometown of Aga-machi—under the shadow of its namesake Mount Kirin—while the other 8 percent is harvested elsewhere in Niigata Prefecture (the goal is to eventually get to 100 percent locally grown). The company doesn't use the most popular variety of sake rice, Yamada Nishiki, because it doesn't grow in Niigata. Talk about commitment.

But it doesn't end there. Each summer, the toji and the kurabito work in the nearby rice fields to cultivate and harvest the rice they'll use during the subsequent fall and winter brewing seasons.

Before founder Kichizaemon Saito officially kicked off his sake-making endeavors in 1843, his family's main business was charcoal—a big industry in Aga-machi a couple hundred years ago. As charcoal-producing activity declined, the Saitos transitioned to making nihonshu full-time—a craft that they've mastered for seven generations.

The family, now led by CEO Shuntaro Saito—who took the reins in 2006 from sixth-generation leader Kichihei Saito—is devoted to preserving not only its nearly 200 years of brewing tradition but also the land they call home. Dense forest covers more than 90 percent of Aga-machi, and Kirinzan is intent on keeping it that way. In 2011, the brewery launched a tree-planting project to give back to the company's natural surroundings.

The Kirinzan portfolio includes plenty of food-friendly options, including its futsu-shu, Karakuchi, and Kirinzan Junmai, both of which the brewery recommends pairing with clams and other bivalves. It suggests everything from toasted marshmallows to pork tenderloin with its junmai ginjo and steamed snapper and potato gnocchi with Kagayaki, its signature daiginjo. For its junmai daiginjo, roast breast of duck with caramelized apples is recommended.

FUSHIMI, KYOTO

Sake brewing in Kyoto's Fushimi district exploded during the early part of the Edo period (1603–1868). Like Kobe's Nada, Kyoto's Fushimi sake district is quite walkable and one of the best ways to spend an afternoon (or three) while in the city. There are around three dozen breweries throughout the neighborhood, including the historic site of one of the largest, Gekkeikan (even those who've tasted sake only a handful of times have likely had one of Gekkeikan's products, as they're the most widely available in the United States, thanks largely to the 1989 establishment of a brewery in northern California). Gekkeikan has been a fixture in Fushimi since 1637, and many of the buildings throughout the neighborhood were once functioning components of the brewery. Among those is the Gekkeikan Okura Sake Museum, which operates inside a former brewing facility built in 1909. There's plenty of antique sake brewery eye candy—old vessels and such—and the tour does its best to recreate the vibe of a less techno-logically advanced period in nihonshu production. You'll even get to taste a few Gekkeikan selections at the end of the tour, and there's also an optional tour that takes visitors to the functioning brewery next door where you can witness some of the process in action.

Despite the significant number of operating sake production facilities in the neighborhood, very few are open to the public for tours. But what you're really here for is the tasting, and when it comes to sampling the wonderful fruits of Kyoto brewers' labor, it's an embarrassment of riches. A good place to start is at Fushimi Yume Hyakushu, which sits inside what once was another of the buildings on Gekkeikan's sprawling campus. It first opened in 1919 as the brewing behemoth's head office; today it's a pretty low-key café and gift shop selling the likes of expertly curated three-cup tasting flights.

If you really want to up the sampling ante, you'll want to head to the decidedly more modern Otesuji shopping arcade (think quasi-open-air mall, with an arched roof to protect shoppers and diners from the elements). Inside the arcade resides one of the best sake educational facilities for the casual consumer, outside the formal schools and sommelier training programs. That would be Ginjo-shubo Aburacho, a modest shop with a gargantuan selection of local sake. Aburacho carries about 80 different bottles from Kyoto breweries, but that's not even the main attraction. There's a tasting bar in the back of the store where you can sample just about any sake Aburacho has in stock in a three-cup setup similar to that on the menu at Fushimi Yume Hyakushu—except there are far more options to choose from to design your own bespoke flight. (I must qualify that by noting that the menu is entirely in Japanese with no English version—as is usually the case in much of Japan outside major tourist areas. However, the owner speaks English, and he will gladly build a flight for you based on your flavor preferences.) Your selections will be accompanied by a couple of small culinary accompaniments: a cube of tofu that serves as a palate cleanser as well as some tasty, umami-rich miso paste mixed with vegetables. I would have tasted my way through Aburacho's entire sake inventory if I had a few hours (and a few extra livers) to spare. But I really wanted to get to the next destination, Torisei, a wonderland for lovers of yakitori (grilled chicken skewers, though the term has come to be applied to just about anything that's cooked on a small stick). Torisei also happens to be the de facto tasting room for the Yamamoto Honke sake brewery, which has been producing in Kyoto since 1667.

The terms *namazake* and *draft sake* are often used interchangeably, but for me, the only time the unpasteurized style lived up to the latter classification was when I sipped the brewery's Kuradashi Nama Genshu while chowing down on skewers of chicken skin and cartilage, as well as some of Kyoto's famous local beef. The servers poured it from a large tank—essentially a keg—which fits the definition of draft much more than bottled nama, at least semantically. For all intents and purposes, it was the "house sake," and if you've ever ordered something listed as the house sake, you know it's not exactly going to be premium. Such a notion went completely out the window at Torisei.

If you are interested in some pre-dinner nihonshu and savory skewers, plan accordingly. My wife and I went to Torisei for a yakitori-and-sake happy hour, and by about 4 p.m., the place is pretty packed.

I always like to stay hydrated when I drink sake (or any other alcohol), and it was a real treat to do so in Kyoto because the water is just so damned good. And the water source is likely to remain pristine as the district doesn't allow underground tunneling for new construction projects.

You don't have to be in a bar or restaurant to enjoy a few sips of the sacred H_2O for which Fushimi is so famous. There are working springs at various points throughout the district from which anyone may partake. It's quite common to witness locals filling plastic bottle after plastic bottle to bring home—not necessarily to drink but to cook with, as it's that much better for food. A walking map is available at major tourist sites so you can easily find all of the springs and keep your thirst quenched. There's one right outside Torisei, as a matter of fact.

And while you're at it, I highly recommend taking a stroll along the network of canals throughout the area. It's quite stunning in November, as that's peak leaf-peeping time in the city—about a month behind the pinnacle of fall foliage splendor in the northern United States, so you can double dip if you're so inclined. For a small fee, you can take a ride on a traditional canal boat—a jukkokubune—but I'd advise against it if you're trying to avoid hyper-touristy activities.

Having said that, one attraction in particular I'd urge anyone not to miss is Fushimi Inari-Taisha, a giant shrine to the Inari, the god of rice and, therefore, one of the primary deities associated with sake production. It's famous for its seemingly never-ending array of torri, orange gate-like structures arranged into a series of tunnels leading to all the major religious structures. It takes a couple of hours to get through and there are some steep hills, but it's well worth it and counts as your workout for the day.

While in Fushimi, I was also fortunate enough to participate in a private tour of Tamanohikari Shuzo. Though the current building is a rather youthful 120 years old, Tamanohikari boasts a 1673 origin story—and the twelfth-generation descendant of the founders now serves as chairman. But it was the prior generation that really put Tamanohikari on the map as far as modern sake making is concerned.

The brewery's claim to fame is that it was the first to resume postwar junmai production in 1964 (brewers added distilled alcohol to the sake mash out of necessity due to rice shortages and rationing immediately before, during, and after World War II). It was a bold move, considering how expensive it was. Junmai uses nearly twice the amount of rice that spirit-enhanced sake does (depending on the grade, of course. If it's junmai daiginjo, it'll actually require a lot more since so much of the rice gets scraped away). The crazier part was that Tamanohikari refused to raise its prices. Nobody else at the time was making junmai, so no other brewery needed to shell out extra cash for their raw materials. A higher price likely would have put Tamanohikari out of business, especially since few drinkers at that time could fully appreciate the allure of a pure-rice nihonshu.

Needless to say, Tamanohikari's gamble ultimately paid off, and to this day, the brewery produces only junmai ginjo– and junmai daiginjo–grade sake—a distinction that the majority of producers can't claim. And today you'd be hard-pressed to find a family-owned brewery in Japan that doesn't brew at least a junmai, junmai ginjo, junmai daiginjo, or all three.

This, of course, doesn't necessarily make Tamanohikari better than those companies that craft honjozo and futsu-shu, along with their junmai offerings. It's usually a stylistic choice rather than an economic one. But every tradition has its purists, and before I knew much about sake, I was one of those purists. But it didn't take long for me to learn that I was really shutting myself off from some genuinely wonderful beverages. I do tip my hat, however, to Tamanohikari because the shuzo made its devotion to sub-60-percent-seimaibuai junmai nihonshu its calling card. And its toji has successfully demonstrated that there's so much flavor diversity, even when so much of the rice is polished away and the mash contains nothing but rice, koji, yeast, and water.

Tamanohikari's owners, like those at Kiminoi and Matsunoi, take great pride in doing all of their rice milling in-house.

"Our chairman has a philosophy: from start to finish, it is all my brewery," says Yuji Fukai, who manages Tamanohikari's overseas sales. Yuji was my liaison at the company's Fushimi facility, which is the familiar mix of old world and modern I'd come to expect from breweries based in historically relevant sake-making areas. There's that rustic wood interior in parts of the brew house that both contrast and complement the stainless steel industrial components of others. Even the kojimuro flip-flops between a classic and a

77

twenty-first-century aesthetic, with cedar boxes used for daiginjo-grade products and stainless steel for everything else.

About 30 unassuming round holes dot the floor of the second story, each the mouth to a large fermenting tank. How large? Each tank holds about 2,000 isshobin—those 1.8-liter bottles—worth of moromi eager to become Tamanohikari's next sake to reach the marketplace. Much of it is also stored at its second facility about 25 kilometers away.

Yuji-san let me taste some of Tamanohikari's Shuhukon junmai ginjo that just went into a bottle—so fresh the container had yet to be labeled. It was a bit sweet, a bit creamy, and there were definitely a few green apple notes. I wasn't familiar with Shuhukon, but Yuji assured me that I was. It's marketed outside of Japan as Tokusen (a much more recognizable name to me). Oh, and by the way, everything within the brewery walls is certified kosher, which certainly gives Tamanohikari an edge in international markets.

When I visited Tamanohikari in late November, it was still relatively early in the sake-making season. From October through January, the operation is junmai-ginjo-exclusive. In February and March, the company shifts its focus to junmai daiginjo (and to the woodier elements of its kojimuro). Production wraps up by April, and for the late spring and summer, most business activities are focused on sales and marketing. Then, in September, the rice is ripe for the picking, and the whole cycle begins anew.

A BRIEF DETOUR

When I first visited Osaka, I remarked, "This is pretty much Japan's Chicago, complete with a narrow river running through the center of it." I didn't realize how spot-on I actually was. Turns out, sometime later, when I was on the subterranean moving walkway at O'Hare International Airport, I noticed a group of banners celebrating Chicago's "sister city," Osaka, Japan. Nailed it.

Osaka is just as vibrant and chaotic as its elder sibling, Tokyo, and those making a sake pilgrimage to Japan should definitely spend a little time there. One of my most cherished travel memories is the time I visited an unassuming little sake shop known as Shimada Shoten. There's nothing remarkable about the store when you first enter—at least that's what I thought, until I asked about the tasting and the shopkeeper motioned to a hidden stairway. I know this sounds like how most horror movies begin, but it's important to have a little faith.

Descend the rickety stairs (watch your step and watch your head!) and you'll enter what appears to be the hideout for a secret society (let's just call it the "Koji Collective"). A vast assortment of cups, carafes, and other sake wares fill the shelves. There's a massive round wooden table, made from what appears to be the lid to an old fermentation vessel, around which an assortment of Japanese and international drinking enthusiasts sip about six different sakes each from around the country for the nominal fee of just ¥220 (about $2) per taste. Visitors can also pair it with chunky miso, umeboshi (pickled plums), or cheese made with ginjo sake lees, each also priced at ¥220. Each pour goes into a different small glass, and each snack is on a different plate. When you're finished, count the number of glasses and plates you've used, go upstairs, and tell the cashier how many you've had.

But before you do, head inside what appears to be a closet but is actually a cellar full of nihonshu selections at which you can marvel. Just about anything you taste that day is available in the first-floor shop. You're not likely to leave the place empty-handed, no matter how hard you try to resist the temptation.

The quaint sake shop with the secluded tasting dungeon is not the only nihonshu-related memory that stands out from my time in Osaka. I'll always remember the city as the place where I first tried hot sake with a blowfish fin steeped in it (not as poisonous as it sounds). And I wasn't even looking for such a thing—I just stumbled behind a red curtain down a dark alleyway and into a pocket-sized restaurant that I didn't even know existed before that moment).

HIROSHIMA

The tourism website Visit Hiroshima says, "Hiroshima is one of Japan's three great sake-producing *areas*, along with Fushimi in Kyoto and Kobe's Nada district." That's a bit misleading. Yes, the Saijo district in Higashihiroshima City (about a 40-minute train ride from the center of Hiroshima City) rivals Nada and Fushimi as a neighborhood with distinctive sake heritage. There are 10 breweries within Saijo (out of about 50 in the entire prefecture), most of which are within relatively short walking distance from each other. But to call it one of three big "sake-producing *areas*" neglects the entire prefecture of Niigata. I will concede, however, that there isn't a single city within Niigata with such a concentration of nihonshu-making activity.

It's pretty spread out, which makes sense since it's a mostly rural, mountainous prefecture.

Today Saijo is most famous within sake circles for the annual Saijo Sake Matsuri, a sprawling festival that takes over an entire park dubbed "Sake Hiroba." Saijo Sake Matsuri usually takes place over two full days in early October, the beginning of the traditional brewing season, and draws more than a quarter of a million people. It works just like a wine or beer festival, where attendees would pay a fee usually equivalent to about $15 or $20 and sample a seemingly endless selection of brands produced in the region. But the fun isn't confined to just the Hiroba. The breweries themselves, most of which line the streets near Saijo's main train station, open their doors and allow revelers to sip their wares. The event is also famous for its "5,000 Person Izakaya," an outdoor izakaya that actually holds that many people (think an al fresco version of the gargantuan beer tents at Oktoberfest—but bigger).

Back in 2008, the Saijo Sake Brewers Association developed a Saijo Sake logo to identify brands produced within the neighborhood known as Sakagura Dori ("sake brewery street"). To qualify as Saijo Sake, the beverage must be made using Hiroshima's traditional three-stage brewing process. Its rice must be 100 percent Hiroshima-grown, and it must use water from wells that the breweries themselves maintain. It must also be of junmai or ginjo grade—actually, one might even argue junmai ginjo or daiginjo grade, as the rice must have a seimaibuai of 50 percent or lower for ginjo (technically daiginjo) and 60 percent or lower for junmai (which is the cutoff for ginjo grade). The sake also must pass a blind taste test by an independent judging panel to ensure it's of the quality worthy of the Saijo Sake name. The best known of the breweries in Saijo is Kamotsuru, which has been there since 1873.

Generally speaking, Hiroshima sake tilts a bit toward the sweeter side, and it also has a great deal more umami character than some of the nihonshu produced in other prominent Japanese brewing prefectures. Its mouthfeel often can be considerably richer than that made in places like Nada or Niigata.

Saito comes close to rivaling Fushimi for the quality of water. And, as is the case in that Kyoto district, folks can taste it in the raw at various springs throughout the area.

·CHAPTER 9·

TASTING NOTES: CURATED SELECTIONS FROM THE BIG FOUR

I included a few bottles from the Big Four sake prefectures in the yamahai and kimoto flight a few chapters back. Here are several other irresistible offerings from that prime nihonshu real estate. But take your time with these. This section includes more samples than others because there are just so damned many of them!

83

Producer: Hakutsuru Shuzo
Prefecture: Hyogo
SMV: +2

I've never experienced a greater sense of cognitive dissonance with a single beverage than I did the first time I tried Hakutsuru Toji-Kan. For that reason, my impressions on this one are going to be a bit longer than others. I was in San Francisco for Sake Summit, part of the J-Pop Summit, an expo celebrating Japanese pop culture, art, and cuisine. As I entered the sake room, a designer masu maker who was one of the sponsors poured samples of Toji-Kan. It kind of knocked my socks off. Robust aromas of cooked rice, roasted nuts, a touch of soy sauce, and maybe a faint suggestion of aged gouda exploded from the cup. That, combined with the rich, creamy texture, made me think, This has got to be a junmai, tokubetsu junmai, or tokubetsu honjozo. (It has a seimaibuai of 70 percent, which would put it in honjozo or junmai territory.) Oddly, its sell sheet included none of that information. When I got to the Hakutsuru table, I finally asked and the rep told me it was a futsu-shu! I was floored. Futsu-shu has such a reputation for being generic and unremarkable. But not this one. When Hakutsuru's toji, Masao Nakazawa, was about to retire, the company asked him to brew a sake that he would like to drink. So he applied his masterful brewing skills to craft a futsu-shu that, at least in my mind, challenges everything everyone thinks they know about futsu-shu. He also chose one of the most prized varieties of sake rice, Yamada Nishiki, to brew the product.

TAMANO HIKARI JUNMAI GINJO

Brewer: Tamano Hikari Shuzo
Prefecture: Kyoto
SMV: +1

This particular junmai ginjo definitely swings more on the nutty/grainy side of things, with plenty of roasted almond and steamed rice to go around. There's still some complexity to be had, with a little bit of pineapple and rustic cheese to keep you interested. It has a soft beginning, but things start to get sharp and spiky pretty quickly, with a moderate to medium finish.

84

TAKUMI JUNMAI GINJO

Brewer: Kyohime Shuzo
Prefecture: Kyoto
SMV: 0

The tropical fruit aroma asserts itself confidently—so much so that you'd be forgiven for thinking that it's a junmai daiginjo. However, there's a definite sharpness to this crisp sake that stabs the tongue intermittently.

EIKUN KOTO SENNEN JUNMAI GINJO

Brewer: Saito Shuzo
Prefecture: Kyoto
SMV: +3

There's lots of lychee and pineapple on the nose, and up front it's a tad sweet on the palate. But that quickly gives way to a mostly dry, sometimes funky, yet always smooth junmai ginjo. The smoothness likely comes from one of Kyoto's legendary waters, Fusui.

SHINSEI JUNMAISHU

Brewer: Yamamoto-Honke
Prefecture: Kyoto

There's a delightful richness to this junmai, with plenty of fruit and whipped cream to go around. I'd almost go as far as to say it's got some "strawberries and cream" character. Almost.

SHINSEI KOSHU

Brewer: Yamamoto-Honke
Prefecture: Kyoto

If you're interested in discovering how crazy-funky koshu, (aged sake) can be, you've come to the right place. It pours a very deep amber, practically brown color, and is exploding with all sorts of savory and umami character. It reminds me of a combination of bone broth, teriyaki, and glazed ham. Much better than it sounds, trust me.

85

MATSU NO MIDORI JUNMAI DAIGINJO

Brewer: Yamamoto-Honke
Prefecture: Kyoto
SMV: +5

As far as junmai daiginjos go, this one is pretty straightforward, with some floral and fruity character. But if I had to pick out a dominant note, I'd say there's definitely licorice happening here.

KIZAKURA JUNMAI GINJO TARUSAKE

Brewer: Kizakura Shuzo
Prefecture: Kyoto
SMV: +1

Kizakura ages this junmai ginjo in Yoshinosugi wood, which imparts a pronounced toasty quality. At times it evokes the charred outer layer of a toasted marshmallow that's touched direct flame, crossed with the caramelized-sugar crust of crème brulée. That being said, would you believe it's actually on the dry side?

KAMOTSURU TOKUBETSU JUNMAI

Brewer: Kamotsuru Shuzo
Prefecture: Hiroshima
SMV: +3

Strong aromas of almonds, steamed rice, yellow apple, and even a hint of pungent, stinky cheese manifest in this tokubetsu junmai. It's medium to full-bodied, and it finishes quite sharply. It's a foodie's dream, as it will hold up against some pretty intense flavors on the dinner table.

HAKUTSURU NISHIKI JUNMAI DAIGINJO

Brewer: Hakutsuru Shuzo
Prefecture: Hyogo
SMV: +4

There's not as much intensity on the nose as you'd expect from a junmai daiginjo, but there's enough melon there—crossed with a touch of steamed rice—to keep you intrigued. The main attraction on this one is its rich and creamy mouthfeel, which was an interesting surprise for me. I don't usually think *mild cheddar mac-n-cheese* when I think junmai daiginjo, but there you have it.

SASAIWAI JUNMAI MUROKA

Brewer: Sasaiwai Shuzo
Prefecture: Niigata
SMV: +3

Sasaiwai Junmai Muroka—"muroka" refers to a lack of filtration—is quite fruity, especially for a junmai (its aromas suggest something in the ginjo or daiginjo wheelhouse). Pineapple and lychee are the dominant notes, creating the perception that more sweetness is present than is actually so (it's mostly dry). Crisp and clean immediately come to mind, thanks to that snowy Niigata water. Yamada Nishiki rice gets a lot of credit for its character as well.

KIKUSUI NO KARAKUCHI

Brewer: Kikusui Shuzo
Prefecture: Niigata
SMV: +7

The aroma is a bit shy, but when it does assert itself, there's some melon and a little bit of sweet rice on the nose. It's crisp, dry, sharp, and quite satisfying over all.

KANBARA "BRIDE OF THE FOX"

Brewer: Kaetsu Shuzo
Prefecture: Niigata
SMV: +3

The 50-percent seimaibuai puts this junmai ginjo on the cusp of junmai daiginjo territory, but aroma-wise there's a lot more going on than the typical floral and fruit notes you'd expect from the latter classification. There's definitely a bit of fruit there, maybe some slightly tart berry or stone fruit, but your nose is also likely to pick up some richer notes of steamed rice and dry roasted nuts. I fell hard the first time I tried this nihonshu on a trip to Portland, Oregon (I mean in love, not on the floor).

KIRINZAN JUNMAI GINJO

Brewer: Kirinzan Shuzo
Prefecture: Niigata
SMV: +4

Definitely more roasty, steamed rice and nutty notes on the nose than fruit, which makes it fairly iconoclastic among junmai ginjos. It's got a somewhat creamy mouthfeel, and some hints of mushroom and butter to boot. Usually I'd say pair a ginjo with an appetizer, but this one's definitely a main dish kind of sipper.

KIRINZAN JUNMAI DAIGINJO

Brewer: Kirinzan Shuzo
Prefecture: Niigata
SMV: +1.5

It's considerably sweeter than Kirinzan's signature junmai ginjo, especially on the finish. On the nose it evokes a cupcake with strawberry icing (sort of).

YUKIKAGE "SNOW SHADOW"

Brewer: Kinshihai Shuzo
Prefecture: Niigata
SMV: +4

This particular tokubetsu junmai is the best seller internationally (but not so much within Japan) for Kinshihai Shuzo, located in the city of Gosen in the Niigata Prefecture. The makers say they brew it for the younger generation of drinkers to pair with many different kinds of non-Japanese food. And it makes sense because it's like a side dish in a glass with a tasty combo of cooked rice, toasted sesame, roasted walnuts, and umami notes.

MURAMATSU GINJO

Brewer: Kinshihai Shuzo
Prefecture: Niigata

Muramatsu is actually the old name for the town of Gosen, where Kinshihai Shuzo produces this grain-forward ginjo. It's got a touch of sweetness and a few intermittent umami expressions, and the brewery's general manager notes that it pairs remarkably well with the traditional carp dishes of Niigata. But fret not, there are hundreds of other meals it goes just as well with if you can't get your hands on any carp.

ECHIGO TOJI JYUN GIN

Brewer: Kinshihai Shuzo
Prefecture: Niigata

As you've probably guessed by the name, this is a junmai ginjo. It's a mostly fruity affair, with hints of citrus and green apple. But there's enough grain, nut, and umami elements to counterbalance the fruit. There's even more of a cereal quality as it approaches room temperature. I could have sworn that I nosed some Grape Nuts–like character even though barley—of which the Post cereal is made—would never be mistaken for rice.

ECHIGO TOJI JUNMAI DAIGINJO

Brewer: Kinshihai Shuzo
Prefecture: Niigata

Like its junmai ginjo sister (when chilled), Echigo Toji Junmai Daiginjo is rather fruit-forward, though maybe in a toasty way. If that doesn't make sense, then think of a grilled pineapple and you're pretty close to the olfactory sensation this one evokes. But overall it's pretty delicate and, dare I say, elegant?

89

ONDA JUNMAI

Brewer: Onda Shuzo
Prefecture: Niigata
SMV: -2

The SMV alerts you that Onda is going to be fairly sweet. It's not cloyingly so, as there are some interesting cheese-like elements to keep it from being too one dimensional. Think three-year-aged gouda crossed with Stilton. Its richness makes it something of an anomaly among Niigata-produced sakes, which tend to be cleaner and lighter. And from a polishing standpoint, it goes above and beyond for a junmai, as it rocks a 48 percent seimaibuai, crossing a couple of points into junmai daiginjo territory.

ECHIGO TSURUKAME WINE YEAST

Brewer: Uehara Shuzo
Prefecture: Niigata
SMV: -32 (for real!)

I don't think I've ever sampled a weirder sake. And that seems to be by design, as the brewer ferments it with wine yeast, which makes it smell and taste much like a dessert wine (at -32, it's unapologetically sweet). It also would be easy to mistake it for a moderately sweet cider (not the bubblegummy, mind-numbingly confection-like sort, but something in the way of a sweeter Angry Orchard offering or a Quebecois ice wine).

ISHIMOTO JUNMAI DAIGINJO

Brewer: Ishimoto Shuzo
Prefecture: Niigata
SMV: +4

No big surprises here. It's light, fruity, and extremely delicate—you know, everything a junmai daiginjo should be. For those keeping score, it has a seimaibuai of 48 percent.

YOSHIKAWA-TOJINOSATO

Brewer: Yoshikawa-tojinosato
Prefecture: Niigata
SMV: -2

It's nearly as complex as its name. It's a bit like a light breakfast with subtle bits of yogurt, cheese, and melon. It's mostly sweet, but the sharp acidity makes it seem drier than it actually is (acidity is kind of like wind chill factor in that regard).

MIDORI KAWA

Brewer: Midorikawa Shuzo
Prefecture: Niigata
SMV: +4

Midori Kawa ("Green River") honjozo features a seimaibuai of 60 percent and a remarkable balance between fruit and toasted grain, with notes of melon, hazelnut, and a bit of stir-fried soba noodles. Still, it's clean and crisp enough to be an ideal companion for sushi.

KIKUSUI JUNMAI GINJO

Brewer: Kikusui Shuzo
Prefecture: Niigata
SMV: +1

The very definition of a no-nonsense junmai ginjo: light and refreshing with aromas of citrus and melon and a silky mouthfeel with a bit of sweetness. Drink it with a salad (or don't).

HAKKAISAN TOKUBETSU HONJOZO

Brewer: Hakkaisan Shuzo
Prefecture: Niigata
SMV: +4

Up front, this one is clean and dry with some tropical fruit and melon. When warmed up, some earthy elements co-mingle with a certain ricey quality. Try it both ways—you'll have a different experience each time.

KUBOTA MANJU

Brewer: Asahi Shuzo
Prefecture: Niigata
SMV: +2

It's quite easy to drink, but that doesn't mean you should drink it easily. There are a lot of delicate nuances in this junmai daiginjo that you don't want to just slurp through. When sipped slowly, you'll probably notice a touch of sour apple and maybe a little bit of cooked rice.

KUBOTA SENJU

Brewer: Asahi Shuzo
Prefecture: Niigata
SMV: +5

Three words come to mind: *dry, dry,* and *dry.* I was caught a bit off guard by how dry it was considering that an SMV of +5, while firmly on the dry side of the scale, would hardly be considered "extra dry" or even "very dry." It's the picture of crispness, with a subdued, incredibly gentle aroma of banana and a few sprinkles of baking spice.

MATSUNOI TOKUBETSU HONJOZO

Brewer: Matsunoi Shuzo
Prefecture: Niigata
SMV: +5

Your nose has to go pretty deep into the glass to detect an aroma, but once you get close enough you might find a little banana bread hiding down there. You'll also want to stick around for the texture, which has a certain richness to it.

DAKU

Brewer: Musashino Shuzo
Prefecture: Niigata
SMV: -12

If I had the opportunity to add a subtitle to Daku, it would be "How I Learned to Stop Worrying and Love Nigori." I've made no secret of the fact that nigori is not my favorite style of sake, mainly because it tends to be a little too far on the sweet side for my taste. However, the junmai-grade Daku turned me around. It was a good thing I didn't know the SMV before I tried it because I would have let my preconceived notions cloud my judgment. It's creamy and rich, just as one would expect from a cloudy sake, but there's also a dryness that balances it all out. It's more perceived dryness than actual dryness, as the moderately high level of acidity can help trick the mind in that way. Try this one with raw oysters—you won't be disappointed.

TEN-TO-CHI "HEAVEN AND EARTH"

Brewer: Musashino Shuzo
Prefecture: Niigata
SMV: +5

"Heaven and Earth" is a nice showcase for Niigata's signature rice breed, Koshi Tanrei, milled just enough (50 percent) to make it a junmai daiginjo. It's decidedly grain-forward, with some light umami and caramelized sugar notes, vaguely reminiscent of a hard, aged cheese, like a three-year-old gouda.

NYOKON "INTO YOUR SOUL" TOKUBETSU HONJOZO

Brewer: Matsunoi Shuzo
Prefecture: Niigata
SMV: +3

It definitely will burrow its way into your soul with notes of melon, plum, and pineapple. The fruity elements come from the use of Kyokai No. 9 yeast, which is combined with Gohyaku Mangoku rice milled to 60 percent and soft, subterranean water, courtesy of the nearby Myoko mountains.

KOIN 1994

Brewer: Musashino Shuzo
Prefecture: Niigata

This isn't one you're likely to find, but I wanted to include it because it illustrates just how amazingly complex an aged sake can be. Musashino Shuzo general manager Hisashi Kobayashi cracked open a bottle of this bad boy when I visited the brewery in late 2017—a full 23 years after its contents were produced. The extended time the beverage spent in its tranquil repose resulted in a mind-blowing symphony of toffee, caramel, candied pecans, and butterscotch that was more evocative of a fine Cognac, sherry, or port than sake.

YOSHINOGAWA DRY SAKE

Brewer: Yoshinogawa Shuzo
Prefecture: Niigata
SMV: +5

As a company, Yoshinogawa is noteworthy because it's the oldest sake brewery in Niigata Prefecture, having begun its journey in 1548. Its dry honjozo is a bit rough around the edges, but it's incredibly lively.

The flavor is all rice, and it's almost a bit chewy at first. There's not much of a nose to speak of, but the ethanol can get a bit assertive on the palate toward the back end.

HIRO SAKE RED

Brewer: Taiyo Shuzo
Prefecture: Niigata
SMV: +4

With notes that sometimes evoke toasted banana nut muffins, Hiro Sake Red junmai is one of those sakes you can flip a coin to decide whether you want to enjoy it hot or cold. It's a flavorful, aromatic experience either way.

KIMINOI

Brewer: Kiminoi Shuzo
Prefecture: Niigata

Kiminoi's eponymous futsu-shu is its most popular product across the Niigata Prefecture. It's a little sweet at first, but then gets kind of exciting (especially for a futsu-shu) with a chewy, almost chalky texture. That may not necessarily sound appealing, but believe me, there wasn't a drop of it I didn't love.

KIMINOI HOUSE WINE

Brewer: Kiminoi Shuzo
Prefecture: Niigata

It's a strange name, to be sure, but House Wine walks the straight and narrow. It's crisp and dry with some slight melon-like characteristics. When you look up *no-nonsense* in the dictionary, this is what you'll likely find.

KIMINOI JUNMAI GINJO MUROKA

Brewer: Kiminoi Shuzo
Prefecture: Niigata
SMV: +2

Kiminoi presents the perfect example of how eliminating activated carbon filtration changes the character of a sake. At first I thought it'd be a little too "inside baseball" for my still-developing palate, but I really did taste the difference when I drank it side by side with the non-muroka junmai ginjo. I've got three words for this one: *banana cream pie.*

MATSUNOI TOKUBETSU JUNMAI

Brewer: Matsunoi Shuzo
Prefecture: Niigata
SMV: +5

I was somewhat confounded by this tokubetsu junmai. It starts off considerably more watery than its honjozo counterpart, but it soon begins to reveal some stealthy complexity, mostly of the mildly funky blue cheese and tropical fruit sort. And that's just when you drink it chilled. When I visited the brewery, Matsunoi Shuzo CEO Makoto Furusawa threw a little of the tokubetsu junmai in a black earthenware pot (in the shuzo's traditional tatami room that beats the hell out of the type of soul-sucking, fluorescent-lit conference rooms with which we're all familiar) and gave it a little fire to demonstrate the night-and-day experience that is drinking it hot versus cold. The heat turns it into a delightful umami bomb right down to the last drop. It also had some notes that were vaguely reminiscent of butterscotch Krimpets.

MATSUNOI JUNMAI GINJO

Brewer: Matsunoi Shuzo
Prefecture: Niigata

It's light on aroma and quite clean, so if you're in the mood for something refreshing, Matsunoi Junmai Ginjo should be in your glass. You'll likely get a whiff of some subtle pineapple, but that's balanced by some unexpected umami that pokes up to the surface like a Whack-a-Mole.

MATSUNOI DAIGINJO

Brewer: Matsunoi Shuzo
Prefecture: Niigata

This softly fruity daiginjo is just so, so elegant. I first tried it at Matsunoi's brewery as company executives plied me with assorted pickles made with sake lees, as well as cheese and macadamia nuts. I almost felt bad eating those things with the shuzo's daiginjo because it really begs to be appreciated on its own.

CHAPTER 10

SAKE-OGRAPHY II: OTHER REGIONS OF NOTE

S ure, Niigata, Kyoto (Fushimi), Hyogo (Nada), and Hiroshima Prefectures are the Big Four, but I never said they necessarily produce the best nihonshu that Japan has to offer. Nor will I attempt to say which prefecture can claim that distinction (take such assertions with a grain of salt, regardless of who is making them). While sakes produced within that quartet of Honshu-based geographic areas undoubtedly would be in contention for some "best of" title, breweries all over the Japanese map would be equally deserving.

YAMAGATA

The southern part of Yamagata Prefecture hugs the northeastern quadrant of Niigata, while its northwestern half ends at the Sea of Japan. To the west, it shares a border with Miyagi, to the south, Fukushima, and to the north, Akita. At last count, there were around 50 active breweries throughout Yamagata. It's as snowy and chilly as you'd expect given its location, you'd likely expect a lot of the sort of light, clean sake for which Niigata is famous. That may be true in some cases, but there was a movement among Yamagata-based breweries a few decades ago to develop a distinctive regional style, in contrast with the Niigata tradition. The "Yamagata style" tilts toward fruity and aromatic. Most recently, the Yamagata Prefecture Sake Brewers Association was so intent on guarding its prefecture's brewing heritage that it sought regional protection for it. And in 2016, the association's efforts paid off when the Japan National Tax Administration granted Yamagata sake a Geographical Indication, recognized by the World Trade Organization. Particularly noteworthy is Yamagata's ginjo-grade sake; the proportion of ginjo to other grades produced in the region is among the highest throughout Japan.

AKITA

Akita is one of the northernmost prefectures on the central main island of Honshu. It's home to nearly 40 nihonshu breweries, and like Niigata, it's well known for its ample snowfall. Akita is two prefectures north of Niigata, with the western coast of Yamagata standing between them.

To the east, Akita shares a border with Iwate, another iconic brewing prefecture. Akita may rank fourth in total sake production, but it's second in sake consumption. In other words, they're keeping more of it for themselves!

Akita winters are long, and generally locals believe the cooler climate the rest of the year provides the optimal environment for producing sake. The brewing season typically runs in the fall and winter months; lower temperatures allow for longer fermentation, which deepens the character of the resulting beverage.

As you can guess, there's a huge sense of local pride in Akita brewing—so much so that those who live there often refer to their prefecture as "bishoukoku" or, loosely, "The Empire of Beautiful Sake."

Among the renowned brands to come out of Akita Prefecture are Manabito from the 330-year-old Hinomaru Jozo, Chokaisan from Tenju Shuzo, and Yuri Masamune from Saiya Shuzo. Like Niigata, Akita is home to a much-revered toji guild, Sannai Toji.

Outside of Japan, the best opportunity to sample the prefecture's well-regarded output is through the New York–based Akita Sake Club. The organization was founded in 2006 to promote not only the prefecture's nihonshu but its food and culture in New York City and surrounding areas. The organization typically hosts two tasting events each year, one in spring and one in fall, where attendees can taste around 50 different selections for an admission fee of around $60.

IWATE

Cross Akita's eastern border and you'll arrive in the Iwate Prefecture, which stretches all the way to the country's east coast. Nearly 30 sake breweries call Iwate home. It may not be able to claim the number of producers that a place like Niigata has, but Iwate more than makes up for the shorter brewery roster in prestige. The local guild, Nanbu Toji, is right up there with Echigo Toji in its prominence.

The Nanbu group is certainly one of the largest, with around 400 certified members working in more than 30 of Japan's 47 prefectures. Nanbu Toji are particularly noteworthy for being pioneers of openness. They were willing to share data and brewing techniques at a time when other guilds were more reluctant to divulge their own trade secrets. The Nanbu tradition dates back to the seventeenth century and rose in esteem over the next couple hundred years as Iwate brewers traveled outside the prefecture and shared their skills with producers in other regions. Visitors to Iwate would be wise to drink locally, as most of the sake produced there is available only within the prefecture's borders.

GIFU

The landlocked Gifu Prefecture in central Honshu, located roughly midway between Tokyo and Kyoto, offers the best of both worlds in terms of sake-friendly topography. The snowcapped Japanese Alps feed the rivers that ultimately provide all of the fresh, unadulterated well water that any sake maker could possibly want, while the southern portion of Gifu is a sprawling plain ideal for cultivating premium rice—especially the local variety, Hidahomare. So it's no surprise that the prefecture's more than 40 breweries—including Watanabe Sake Brewery Co., Miyozakura Jozo Co., and Iwamura, among many others—continue to thrive. Gifu also has the other critical element of sake producing—the microbes—covered as well. The Gifu Sake Brewers Association and the Industrial Technology Center developed the G-Yeast strains, mutated from the Kyokai No. 9 yeast. Known for its fruity aromatics, G-Yeast has been on the market since 1997 and has been popular for making junmai and ginjo-grade sake.

Oh, and if you're ever in doubt about what to pair Gifu sake with, a good place to start might be Hida-gyu, the regional beef that gives even Kobe's famed meat a run for its money. Marvel at the gorgeous marbling on a slice of Hida-gyu.

AOMORI

Aomori is the northernmost prefecture on the central island of Honshu, and like some of the other prime sake regions to its immediate south—crossing its southern border gets you to Iwate and Akita—there's no shortage of snow in the winter. The prefecture is home to some 40-odd sake breweries, including Hachinohe Shuzo, the most prominent of those. Based in the city of the same name, Hachinohe

99

has been in operation since 1775 and is best known for its flagship product, Mutsu Otokoyama.

NAGANO

Known for its cool climate—it was the site of the 1998 Winter Olympics, after all—Nagano has risen in prominence as a sake-producing region. And that's quite an accomplishment, considering that it's pretty easy to get lost in the shadow of its northern neighbor, Niigata. There are roughly 80 breweries in Nagano, whose products benefit from a confluence of earthly influences: nippy winds blowing down from the Japanese Alps, immaculate waters cascading down from said peaks, and sprawling rice fields. It's ground zero for Miyama Nishiki, which helps give Nagano sake its famously crisp character.

However, arguably Nagano's greatest contribution to the Japanese brewing industry was on the microbial level. You'll recall that it was at a Nagano brewery, Miyasaka Shuzo, where, in 1946, researchers and producers first isolated Kyokai No. 7, the most popular yeast strain across the entire industry. And 45 years later, the prefecture built upon that legacy with the introduction of the low-acid Alps Yeast— ideal for brewers looking to enhance the fruity and floral aromatics of their products. Miyazaka today remains a beloved producer, especially for the fruit-forward Masumi Arabashiri Junmai Genshu Nama. It just screams "fresh."

FUKUSHIMA

Fukushima Prefecture borders Niigata to the east (and extends to Japan's east coast), so it's no surprise that there's a variety of great sake to be found there.

Unfortunately, the 60-plus breweries in the prefecture have been dealing with a number of significant setbacks since the 2011 Fukushima earthquake and subsequent tsunami and the meltdown of three nuclear reactors near the coast. A handful of operations were completely destroyed, and the ones that weren't have had to constantly work to quell public fears about potential product contamination— despite the fact that most of the prefecture's famous rice fields are located far inland and nowhere near the site of the disaster. Producers reported that they were regularly fielding calls from nervous consumers inquiring about the safety of their beverages.

The events also damaged exports for Fukushima-produced products. For instance, China had banned sake imports from Fukushima and surrounding regions. Additionally, even though there was no explicit governmental prohibition, sales to South Korea—one of the largest export markets for sake—had ground to a halt. And Bloomberg reported that as of early 2018, a total of 55 countries throughout the world have some restriction or requirement for additional documentation on imports of products manufactured in Fukushima.

It didn't seem to matter to those countries, or to many consumers throughout the rest of Japan, that the government fastidiously tests all of the raw materials going into any commercial product, and if any rice or water shows an unsafe level of radiation, it doesn't make it into the mash. It certainly helps that rice growers in the region cultivate their crops with potassium fertilizer, which has been known to produce grain that doesn't absorb radiation.

And since 2015, none of Fukushima's rice has registered radiation above the safety level—a relatively costly effort that the prefectural government funds and the owner of the nuclear power plants subsidizes. The program is very transparent—every bag of rice that's been tested gets its own unique barcode, and anyone with internet access can access the radiation test results for every bag online.

If there's been one silver lining, it's that this has given local brewers the chance to double down on efforts to produce premium-grade products. And that's already started to pay dividends. In the years following the disaster, Fukushima sake makers have been on a winning streak at Japanese and international sake competitions, racking up gold medal after gold medal (and taking home the most gold out of any prefecture each year from 2012 to 2017 at the Japan Sake Awards—not bad for a place that hadn't won any just two decades prior).

The Fukushima Prefecture Sake Brewers Cooperative has begun to promote that fact, helping to calm lingering fears about the region's nihonshu. Meanwhile, the Fukushima Prefectural Government has been allocating the equivalent of nearly $1 million a year on such promotional efforts, both in the home country and abroad.

For one such event held in October 2017, the Fukushima government joined forces with the British-Japanese Parliamentary Group to treat British MPs and other public officials—as well as folks in the corporate sector from both countries—to a VIP tasting of Fukushima sake.

101

The campaign is definitely working, at least internationally. Between 2012—the first full year after the devastating incident—and 2017, Fukushima sake exports more than doubled. The United States has been the largest market for the prefecture's beverage output.

But even before the 2011 disaster, Fukushima brewers had been actively focusing on improving product quality—not to mention the perception of said quality. In 1992, the government set up a sake academy to help develop the skills of existing and future brewing professionals. Since then, the number of premium products coming out of the prefecture increased exponentially, as did the general public's esteem for Fukushima nihonshu.

Thanks in part to geography, Fukushima sake is one of stylistic diversity. The prefecture is home to three distinct regions divided by mountain ranges with varying climates and natural resources, and each has developed its own sake-making tradition. The westernmost portion, the Aizu region, is snowy and mountainous—not surprising since it borders Niigata.

The Nakadori region, which lies to the immediate east of Aizu, is essentially the heartland, where most of the agricultural activity—including rice cultivation—is happening. Finally, the easternmost slice, Hamadori, is the coastal region, with the mildest, most temperate climate of the three.

Suehiro Shuzo, founded in 1850 and now in its eighth generation of family ownership, is one of the bigger names within the Aizu sake scene, known particularly for its mastery of the yamahai method. It's a popular site for travelers, offering hourly guided tours and a café that serves desserts made with the brewery's beverages.

The Nakadori region's Daishichi Brewery predates Suehiro by nearly 100 years and is fiercely loyal to many of the old ways—particularly the kimoto jikomi method, which very few breweries are even bothering with these days. And much of it lacks the funkiness one would come to expect from a kimoto. The acidity and creaminess are still there, but in a very elegant, understated presentation.

SHIZUOKA

Shizuoka is one of the two prefectures—the other being Yamanashi—that gets to claim Mount Fuji. Japan's highest peak gets a great deal of credit for the quality of the sake produced in the region, which sits on the east coast of central Honshu. The famous Fukuryusui water variety derives from the mountain's melted snow, which seeps into the volcanic soil and becomes one of the country's most pristine springs. That soil acts as a natural filtering agent (go, carbon!), which means a nihonshu from any of the many breweries in Shizuoka is always an enchanting experience.

Oomuraya Shuzo in Shimada City is another notable producer in the prefecture, probably best known for the ultra-smooth junmai daiginjo, Wakatake Onikoroshi "Demon Slayer," a complex yet delicate brew that's a prime representation of the prefecture's signature character. Shizuoka sake is known to go quite well with fish, which is no surprise given its proximity to the Pacific Ocean. If you've fallen in love with bonito sushi (as I have), you probably can thank Shizuoka for that. The prefecture is the bonito fishing capital of Japan.

GUNMA

Landlocked Gunma shares borders with Niigata, Fukushima, Nagano, and three other prefectures, so it's easy for the region to get lost among so many storied sake-making centers. But Gunma's producers are intent on showcasing their individuality, which has, rightly, given them a place at the table.

In 2012, 15 brewery members of the Gunma Prefectural Brewers Cooperative Society began marketing a hyper-local variety of sake known as Maikaze, named for the new breed of rice developed at the Gunma Agricultural Technology Center. The recipe also incorporates Gunma Kaze yeast, developed at the Gunma Industrial Technology Center. The breweries using the locally grown, sourced, and cultivated ingredients get to affix a special blue, yellow, and white seal to their bottles, signifying that 100 percent of their contents are native to Gunma.

FUKUOKA

Since Fukuoka is on the island Kyushu, you wouldn't be wrong to suggest that the prefecture's shochu distilling activity grabs much of the spotlight. Fukuoka isn't as closely associated with a single style of shochu as some other Kyushu prefectures like Kagoshima—known for sweet potato shochu—and Kumamoto—famous for rice shochu.

103

Barley boasts a plurality among popular shochu bases in Fukuoka, but distillers there also make a fair amount of the spirit from sweet potatoes, rice, and other bases. Fukuoka happens to be one of the southern island's more democratic prefectures, alcoholically speaking, and nihonshu production continues to be quite robust. There are nearly 60 breweries in Fukuoka, producing sake of varying character. If one word could sum up any sort of overarching flavor theme, it would likely be *ricey*. That's not necessarily in an over-the-top way. Typically the aroma isn't particularly pronounced, settling somewhere in the so-shu realm.

KUMAMOTO

I want to throw in one more Kyushu prefecture for good measure. Like Fukuoka, Kumamoto is much more revered for its shochu-making scene. However, since rice is the dominant base used in Kumamoto, you can bet some of that grain is foregoing distillation and living out its life as nihonshu. And what Kumamoto lacks in quantity— there are only a dozen or so sake makers in the prefecture—it more than makes up for in quality.

The Aso and Kyushu mountains deserve much of the credit for the fine liquid base of the prefecture's sake, as do the Midori, Shira, and Kikuchi Rivers flowing from those peaks. And let's not forget, without the Institute for Kumamoto Brewing, there would be no Kyokai No. 9 yeast strain, and many ginjos and daiginjos just wouldn't be the same.

HOKKAIDO

Since I've mentioned Kyushu, I have to at least give a small tip of the hat to the northernmost major island in the Japanese archipelago. Being so far north, Hokkaido is, of course, quite cold and snowy. And yes, that means great water. That great water also helps irrigate even greater rice—but that doesn't mean it's easy. Researchers and growers have had to learn to adapt the grain to the frosty climate. Not too long ago, Hokkaido-grown rice didn't have the greatest reputation throughout the rest of the country, especially where sake brewing was concerned. Most breweries had to rely on rice cultivated on Honshu. That started to change with the introduction of the Ginpu, Hokkaido's very own premium sake rice variety. The tipping point was 2003, when a Hokkaido brewer won a gold medal for a nihonshu made with Ginpu.

Ginpu became such a hit that when Norwegian craft beer brewer Nøgne Ø decided to venture into sake making, it chose Ginpu for its products (Norway and Hokkaido do have a few climatological elements in common).

Today there are about 15 active sake breweries in Hokkaido, the oldest of which is Chitosetsuru, opened in 1872 in the island's largest city, Sapporo. Other venerable producers include Kunimare Shuzo, launched 10 years after Chitosetsuru and today known as the country's northernmost shuzo; Otokoyama, founded in 1887; Tanaka Shuzo and Takasago Shuzo, which both opened in 1899. Takasago, in 1926, became the first Hokkaido brewer to win a gold medal in a national competition.

ONE LAST NOTE ON GEOGRAPHY . . .

Now, I don't want to burst your bubble, but by the time you read this, there may be fewer players within each sake-making prefecture. Though the beverage and Japan are so inextricably linked, consumption in nihonshu's home country continues to fall incrementally each year. Beer, whiskey, shochu, and other beverage categories both domestic and imported have divided consumers' attention. Right now sake's market share within Japan is less than 7 percent. And that number too will likely be lower by the time you reach this page. More than ever before, sake breweries are relying on international markets to stay in business. And we US-based imbibers represent one of the largest markets, as well as one of the areas of greatest untapped opportunity for nihonshu. So please keep buying the stuff so these sake-making centers are able to continue to thrive.

105

CHAPTER 11
TASTING NOTES:
A TOUR BEYOND THE FOUR

I t's not all about Niigata, Hiroshima, Kyoto, and Hyogo. Let's take a little journey through some of the other notable sake-making prefectures.

TAKENOTSUYU JUNMAI

Brewer: Takenotsuyu Shuzo
Prefecture: Yamagata

Its seimaibuai of 60 percent could put it into junmai ginjo territory, if the brewer were so inclined, and it definitely exhibits some of that grade's signature characteristics. Its aroma leans toward the fruity and floral, with pear being the dominant note. A bit of cherry does sneak through every so often, though, as does a rogue honeysuckle element. All this adds up to solid refreshment.

TAMURA YAMASAKE 4 JUNMAI GINJO

Brewer: Tamura Shuzo
Prefecture: Tokyo
SMV: +5

With a seimaibuai of 55 percent, Yamagata rice, Tama River water and four types of yeast, Tamura Yamasake 4 Junmai Ginjo really brings the fruit. It's got an intensely tropical nose, with notes of pineapple, kiwi, and a little bit of banana.

TAISETSU JUNMAI GINJO "GARDEN OF THE DIVINE"

Brewer: Takasago Shuzo
Prefecture: Hokkaido
SMV: +3

Hokkaido's known for its ample snowfall, and Takasago Shuzo takes full advantage of that by aging its junmai ginjo in igloos. There's a lot of tropical fruit on the nose, but a grainy, toasted-cashew-like character asserts itself quite enthusiastically. The nutty/fruity balance makes me think of trail mix.

TATENOKAWA NAKADORI 50 JUNMAI DAIGINJO

Brewer: Tatenokawa Shuzo
Prefecture: Yamagata
SMV: -2

Pear, peach, and other stone fruits dominate the relatively subdued nose, making way for a generally smooth junmai daiginjo (the "50" is the seimaibuai, so it walks the line between daiginjo and ginjo territory). For those who prefer sweet to dry, this is the junmai daiginjo for you.

TATENOKAWA 33

Brewer: Tatenokawa Shuzo
Prefecture: Yamagata
SMV: -3 to -2

If sake had terroir (and I'd argue that it does), Tatenokawa 33 (aka "3 Peaks") and many other Tatenokawa Shuzo would be among the best tastes of the Yamagata Prefecture you can get. It uses the prefecture's famous Dewasansan rice for this tropical-fruit-forward junmai daiginjo (the "33" signifies its seimaibuai) that's sweet up front but a bit dry on the back end.

108

TATENOKAWA 18

Brewer: Tatenokawa Shuzo
Prefecture: Yamagata
SMV: -6

If you want to see what it's like to test the limits on rice polishing, Tatenokawa 18 (the seimaibuai) is a good place to start. Highly aromatic with notes of pineapple and lychee and a touch of wild funk, this sweet and delicate sake needs to be sampled on its own without any food getting in the way of the experience.

OZE NO YUKIDOKE OHKARAKUCHI JUNMAI

Brewer: Ryujin Shuzo
Prefecture: Gunma
SMV: +10

The nose of this junmai (which skirts the junmai ginjo realm with a seimaibuai of 60 percent) is mostly stone fruit, with cherry notes playing a starring role. When I first tried it, the pourer told me it was extra dry. It didn't seem far beyond medium-dry at first, but then that super crisp, lingering dry finish hit me.

109

MIZUBASHO PURE (SPARKLING)

Brewer: Nagai Shuzo
Prefecture: Gunma
SMV: +4

This is what I plan to drink at midnight on New Year's Eve for the foreseeable future. With the tiny bubbles you'd expect from one of France's most famous exports, this is the Champagne-drinker's sparkling sake. It's produced in the same style as Champagne, with carbonation resulting from secondary, in-bottle fermentation. It seems much drier than its +4 SMV; bubbles can do that sometimes. It also looks a lot more like Champagne than sake, with its yellowish-to-pale-greenish hue (so your party mates on December 31 will be none the wiser). Plus, there's a delightful bitterness on the finish.

KYOKUSEN JUNMAI DAIGINJO

Brewer: Asabiraki Shuzo
Prefecture: Iwate
SMV: +1

Kyokusen Junmai Daiginjo is the sake that reminds me to avoid being too influenced by marketing sell sheets. The importer's tasting notes say there are notes of melon, but I get more pineapple than that. It's just further proof that tasting is one of the most subjective activities there is (so don't take my word for it either!). Aside from that, it is nice and delicate (the seimaibuai is 40 percent). Drink it on its own before a meal or with some very light, chilled appetizers.

CHIGONOIWA KARAKUCHI JUNMAI-SHU

Brewer: Chigonoiwa Shuzo
Prefecture: Gifu
SMV: +13

Yes, it's dry. But it's also a bit of an iconoclast with flavor notes that are both roasty and sour—I'd go so far as to say that the nose is somewhat reminiscent of red cabbage. I was completely captivated and enthralled!

MUKASHI-NO-MANMA

Brewer: Yamada Shuzo
Prefecture: Gifu
SMV: +1

The name means "unchanged by time," and Yamada Shoten makes that moniker count. The shuzo has been using the same classic methods to brew it as the Yamada family's ancestors did in the nineteenth century. Mukashi-no-Manma is a junmai whose rice, the famed local variety, Hida Homare, is milled well into ginjo territory—55 percent, to be precise. Ultimately, it's soft, somewhat earthy, with a definite banana-like element on the nose.

KOZAEMON JUNMAI DAIGINJO

Brewer: Nakashima Shuzo
Prefecture: Gifu
SMV: +2

"Elegant" doesn't even begin to describe Kozaemon Junmai Daiginjo. Pineapple and other tropical fruits dominate the aroma of this Aiyama- and Yamadanishiki-rice-based nihonshu, sporting a seimaibuai of 40 percent and a healthy acidity that helps it hold up to some fattier sashimi like o-toro.

CODY'S SAKE JUNMAI GINJO

Brewer: Watanabe Shuzo
Prefecture: Gifu
SMV: +3

You'd be right if you thought the name sounded a bit Western. It was brewed to commemorate American brewmaster Darryl Cody's tenth anniversary as toji. There's the slightest touch of sweet rice on the nose, balanced by a little bit of stone fruit. It's delicate enough to flirt with the daiginjo realm, though it's not quite there, with its seimaibuai at the exact midpoint between ginjo and daiginjo (55 percent). And true to its Gifu roots, it uses the local Hida Homare rice.

111

HOURAI NINJA JUNMAI-SHU

Brewer: Watanabe Shuzo
Prefecture: Gifu
SMV: +4

This may be the first time I've ever used the term *jazzy* to describe any beverage, let alone sake. But there's just no other word for it. It's bitter, it's mildly sweet, it's acidic, and even, dare I say, a little bit "dirty."

DAISHINSHU KARAKUCHI TOKUBETSU JUNMAI

Brewer: Daishinshu Shuzo
Prefecture: Nagano
SMV: +8

What makes this dry tokubetsu junmai so tokubetsu is the fact that it uses a somewhat rare rice variety known as hitogokochi. The aromatics tilt toward the floral and herbal, with the slightest suggestion of anise.

HOKUSOUSANYU

Brewer: Washi No O?
Prefecture: Iwate
SMV: +2.17 (How's that for specific?)

There's really no better word than *ricey* to describe grain-forward sakes such as this. It's also quite rich and creamy, and I would even go so far as to say it exhibits some custard-like elements. There's a nice balance mid-palate, though the finish is undeniably sweet.

NANBU BIJIN DAIGINJO

Brewer: Nanbu Bijin
Prefecture: Iwate
SMV: +3

Floral and tropically fruity—lychee comes to mind—on the nose, Nanbu Bijin's ("Southern Beauty") daiginjo is surprisingly aggressive on the palate. And that pleases me.

NANBU BIJIN TOKUBETSU JUNMAI

Brewer: Nanbu Bijin
Prefecture: Iwate
SMV: +5

Clean, moderately sharp, and fruit-forward (a little bit of cooked rice pokes out every now and again), there's a crisp, refreshing quality here that's wonderful on a spring day. It's a seimaibuai of 55 percent, which makes it more of a marketing quirk that Nanbu Bijin didn't classify it as a junmai ginjo.

WAKATAKE ONIKOROSHI "DEMON SLAYER"

Brewer: Oomuraya Shuzo
Prefecture: Shizuoka
SMV: +3

As smooth as an ultra-high thread-count bed sheet, this warrior is ready to destroy any demonic entity, with its floral and fruity notes showcasing the likes of melon, a little banana, and even a bit of anise. You'll also detect some sweet rice, which amps up the complexity.

MICHINOKU

Brewer: Hachinohe Shuzo
Prefecture: Aomori
SMV: +14

Super dry and crisp (if the SMV didn't already tell you as much), the tokubetsu junmai is a pretty straightforward sip, with aromas of fresh honeydew, apple, and apricot, a soft, round texture, and a medium finish.

113

CHIYONOSONO "SHARED PROMISE"

Brewer: Chiyonosono Shuzo
Prefecture: Kumamoto
SMV: +3.5

This is a junmai that does, indeed, live up to its promise. Up front it's super creamy and even a bit chewy with a touch of sweetness. But then the toasty, roasty, stir-fried-noodles-and-nut notes start to kick in all of their splendor. There's also a shot of pear and citrus that asserts itself every so often. And then it finishes with unmistakable burned sugar, burnt-to-a-crisp marshmallow elements that found out this strikingly complex pure rice sake knows how to have fun.

AT YOUR SERVICE

S o how exactly do you drink this stuff? The answer depends on how much time you have. Different grades and styles demand different temperatures, and you'd be hard-pressed to find a bar or restaurant that gets it 100 percent right 100 percent of the time. Sometimes it's enough just to know whether a nihonshu should be served hot or cold.

First, it's important to know exactly how cold is cold. Under no circumstances should sake, regardless of style or grade, be chilled to below 41 degrees Fahrenheit (5 degrees Celsius). As for sakes that are meant to drink warm, you rarely want to drink them at temperatures higher than 131 degrees Fahrenheit (55 degrees Celsius). See the box for specific temperatures and their rather descriptive sake-related names.

Broadly speaking, a sake with delicate flavor and aroma nuances doesn't do too well at high temperatures. So, as a rule (but not as an absolute!), you really want to avoid drinking a ginjo or daiginjo that's any warmer than 50 degrees Fahrenheit—the sweet spot is somewhere between "snow-chilled" and chilly blooming season (see box for exact ranges).

Also, if you read a flavor note for a particular brand (either in this book, on a bottle, or on a website) that describes the aroma as "fresh" or "refreshing," usually you'll want to drink that one chilled.

If you are in a situation where you aren't entirely sure what kind of sake is in your glass (as often happens at a non-English-speaking izakaya in Japan where you can't make heads or tails of the label), you

115

rarely can go wrong if it's medium-cold to room temperature. All sakes can be consumed cold; not all sakes should be consumed warm.

That's not to say there aren't some styles that benefit from a little bit of warmth. In fact, their flavor really opens up if their bottle spends some time in a hot bath. The amino and lactic acids respond well to heat. Some that reside on the jun-shu end of the flavor spectrum—a particular tokubetsu junmai, kimoto, and yamahai, for instance—are the most pleasing at around 104 degrees Fahrenheit (aka "low warm"). Also, grades with added alcohol—honjozo or tokubetsu honjozo, for example—hold up better at even higher temperatures of around 113 degrees Fahrenheit.

And then there are some that are equally excellent when a bit chilled or a bit heated—but not necessarily at room temperature. Sometimes figuring that out can make one's head spin.

But you, as a drinker, will rarely have to worry about such precision. Leave that to the servers and sommeliers to figure out (though, if they bring you a scalding hot junmai daiginjo, chances are you're in the wrong place). Usually the best place to look to get some sense of the serving temperature is the label of the bottle itself. And even then it's unlikely that its recommendation will be down to the degree.

One trap you definitely don't want to fall into when trying to decide whether you should heat or chill your sake is trying to apply a single rule to a single style. While you usually can say that, in 99 out of 100 cases, you're not going to want to heat a junmai daiginjo, you can't make a similar blanket statement about, say, a junmai proper.

It really becomes a matter of trusting your own tongue. Is there an intensely pronounced umami element—and by that I mean, does it dominate all other flavor nuances? If the answer is yes and you'd like it to be even more pronounced, go ahead and warm it up to about 104 degrees Fahrenheit (40 degrees Celsius). Heat has a tendency to enhance umami. It has a similar effect on sweet sakes; but if you want it to seem even sweeter, let it sit in the warm bath for a bit.

The level and type of acidity should dictate whether or not to heat a sake. The beverage contains varying levels of different acids, each of which behaves in its own way to changes in temperature.

Generally, sake has high levels of succinic acid, which is also commonly found in shellfish. The beverage typically boasts more than four times the level of succinic acid found in wine, which is why sommeliers frequently have trouble finding the right wine to go with shellfish. It's just not a good match. Nihonshu, on the other hand, can be an ideal pairing. Umami-rich succinic acid becomes more perceptible when the sake is warmed up.

Heating also accentuates the flavor of lactic acid, which helps when pairing with dairy and fermented foods. Keep in mind, though, that raising the temperature also can sharpen the sensation from the alcohol.

On the flip side, sake contains malic acid, the dominant form found in apples. Chilling enhances those brisk, apple-like components. There's usually a much lower concentration of citric and acetic acids— the former of course being the dominant component of lemons, limes, and oranges, and the latter, the core element of vinegar—but those behave better when chilled.

When all else fails, try room temperature. When you're tasting a sake for the first time—tasting, not settling in with an entire glass or carafe—it's good to get a baseline at an ambient temperature. Certain flavor and aroma elements might disappear when chilled, others might recede when heated, but most will likely express themselves in some way at a more neutral temperature of about 70 degrees Fahrenheit.

117

THERE ARE OFFICIALLY NINE LEVELS OF COLD AND WARMTH IN THE REALM OF SAKE TEMPERATURES:

▶→ 41 DEGREES F (5 degrees C): Snow-chilled
▶→ 50 DEGREES F (10 degrees C): Chilly blooming season
▶→ 59 DEGREES F (15 degrees C): Refreshing cold
▶→ 86 DEGREES F (30 degrees C): Sunny ambient warm
▶→ 95 DEGREES F (35 degrees C): Lukewarm
▶→ 104 DEGREES F (40 degrees C): Low warm
▶→ 113 DEGREES F (45 degrees C): High warm
▶→ 122 DEGREES F (50 degrees C): Hot
▶→ 131-PLUS DEGREES F (55-plus degrees C): Very hot

JEFF CIOLETTI

SERVING VESSELS

I've always been a sucker for romance and ritual, especially when it comes to imbibing. That's one of the key reasons I fell head over heels for sake when I first started to explore it in earnest.

I'm a strong advocate of Western-style eateries and watering holes incorporating nihonshu into their beverage programs, as it's very versatile and can satisfy a broad spectrum of drinking occasions and complement a stunningly diverse array of global cuisines. But I also feel that proprietors of such establishments needn't stress about stocking up on the appropriate consumption vessels. If they have a good supply of wine glasses, they're usually good to go.

In fact, a white wine glass is the best vessel in which to serve daiginjo- and junmai daiginjo-grade sakes, as well as many ginjo offerings. You don't want to drink them out of a small sake cup because those don't allow you to experience the full aroma. You need a little distance between your nose and the liquid, just as you would with a glass of wine. You'd only be getting half of the fruity, floral olfactory intensity you'd get from a wine glass.

But it's not like you're always going to be drinking daiginjo, and that's where things start to get really creative.

First, let's address those aforementioned small cups. There are actually many distinct varieties of such vessels, and some are specific to particular regions. Two of the most notable kinds are choko and guinomi. The traditional choko is made of earthenware (though it's commonly available in lacquer, glass, and metal) and is the smaller of the two. It's most commonly used for warm sake, though there's absolutely nothing wrong with drinking chilled nihonshu out of it. A guinomi is a bit bigger and is customarily used for cold sake.

It's always a treat when servers at sake-centric restaurants arrive at the table with a tray full of chokos of all colors and materials. Before you even make up your mind about what to order, you've got to make the all-important decision regarding which cup you want. The first time I experienced this, I was a bit floored. I had bought a number of sake sets, each consisting of a carafe and matching chokos, because I thought everything was supposed to be uniform. But mixing it up is so much more fun—plus there's a much smaller chance that someone will pilfer your cup.

Okay, get ready for this one: janomeno kikijoko. It's such a mouthful of a moniker for such a small cup. It's really a variation of the choko (though frequently guinomi-sized as well), but with a very specific purpose: it's the go-to vessel that brewers and sommeliers use for tasting. Two blue concentric circles sit at the base of the cup (it looks like a snake's eye—the root word, *janome*, actually means "snake eye"). The circles aid in color and clarity analysis.

One of my personal favorites is the masu, a square box made of wood or lacquer that holds a 180-milliliter pour of sake. There's a certain significance to that number. It's the standard measure for a single serving of rice. There were so many such boxes lying around, people started using them for drinking sake.

These days it's quite common to encounter a particular ritual associated with the masu. Servers will place a small glass—sometimes contoured with the mouth flaring outward and sometimes more akin to a generic orange juice vessel—into the masu. They will then pour nihonshu into the glass, past the brim, until it overflows into the masu.

The first time I saw someone do this, I thought the server got distracted and was making a mistake. But no, that's part of the ceremony. No sake brewer or server is 100 percent clear on how this ritual emerged. The prevailing theory is that it is meant as a display of gratitude and generosity. Sort of a literal "may your cup runneth over" kind of thing. The Japanese term for the practice is *sosogi-koboshi*, which combines the words for "to pour" (*sosogu*) and "spill over" (*kobosu*). Some may also use the term *mokkiri sake*. *Mokkiri* is a word usually associated with food served in over-generous portions.

The ritual isn't exactly unique to masu anymore. Often the server will pour sake into a glass that overflows into a saucer below. I don't really blame them for going the less-traditional route—small plates are far easier to wash and store than square boxes. The masu itself frequently is still used as a drinking cup. In that situation, it's placed on top of the saucer.

I once went to a New York tasting of over 50 sakes from the Akita Prefecture with my own personal masu in tow. At least four different servers made a comment along the lines of "Oh, a masu!" I couldn't tell whether they were impressed or mocking me for bringing the wrong type of vessel to the event (I was a novice at that time, and frankly, I didn't even know it was called a "masu" until that evening!). I found personal vindication many years later in San Francisco at the J-Pop

119

Summit. The showcase of Japanese pop culture, entertainment, gaming, and cuisine featured a sake tasting area. At the entrance to that area was a hand-sketched picture of the wooden box with the words "Drinking sake in a 'MASU' [caps and quotes theirs] is Japanese cool style." So I guess I was cool all along.

I later learned that the masu typically had made few appearances outside of major celebrations like weddings. In such settings, masus are stacked into a pyramid and sake is poured in the top vessel, which overflows and cascades down into the lower cups—very similar to the way Champagne is poured at chichi soirees.

If you want to get super fancy when you're drinking sake, do so out of kiriko, stunningly beautiful crystal ware emblazoned with hand-cut designs. Tokyo and Kagoshima are equally famous for their local kiriko, Edo Kiriko and Satsuma Kiriko, respectively.

With something of the so-shu variety—the light aroma, refreshing flavor sort—you have the most flexibility when it comes to choosing a vessel. A choko would suffice, as would a kiriko (if you're a high-roller). And while you're at it, why not let your masu runneth over.

120

The size of the vessel is pretty important, especially since you'll be drinking a lot of so-shu sakes chilled. That's why a choko is probably a better bet than a guinomi, because it's smaller. Since it takes a bit longer to empty the guinomi, the temperature might rise to a sub-optimal level before you're ready for the next serving.

If it's sparkling sake, Champagne rules apply. Drink out of a flute to ensure that the carbonation doesn't dissipate too quickly.

A Bordeaux wine glass might enhance the experience of drinking a jun-shu, as the wider shape helps bring out the rich texture and nutty, cooked-rice flavors. If you're consuming it warm, as is common with many of this type, go with a guinomi. Finally, as juku-shu sake tends to have an amber color, you're going to want to make sure you're drinking it out of a transparent container. A brandy glass works perfectly here, as it not only showcases the golden hue, but it also does wonders for the complex nose.

If you lack any of the recommended glass or earthenware, don't sweat it. These are more guidelines than rules. But if there's one suggestion that I'd be pretty adamant about, it's that you really shouldn't drink nigori out of clear glassware. The rice sediment it leaves on the sides of the glass is kind of unsightly. But if you don't mind that, I'd say go for it.

CARAFES AND KETTLES

The vessels out of which you pour sake can be just as important as the container into which it's being poured. Strongly resembling a tea kettle, the choshi has been used for centuries. The older ones were frequently made of cast iron (many still are), but you're more likely to find ceramic versions. It's common to pour warm sake from it, but it can be used to dispense cold as well.

The serving vessel I've encountered the most is a sake carafe known as a tokkuri. There are two major forms the tokkuri is likely to take. The first one, made from bamboo (or sometimes from a synthetic material designed to look like bamboo), resembles a giant piece of penne pasta that's been broken in half.

The other version has a very bulbous body and narrow neck that flares out at the mouth. They're usually ceramic, but there are many clear glass iterations as well (I own one). One of the best ways to chill room-temperature sake quickly (the fridge can take around two hours) is to immerse the tokkuri up to its neck in ice water. If you're serving warm sake, do the same in a bowl of water that's heated to about 80 degrees Celsius (176 degrees Fahrenheit). Within about three minutes that should bring the temperature of the sake in the tokkuri up to the low to mid-40s Celsius (or around 104 to 113 degrees Fahrenheit). (Under no circumstances, try to heat up sake in the microwave. It's virtually impossible to control the increase in temperature, and you'll more than likely experience some unpleasant olfactory sensations, the result of heating alcohol too quickly.)

Then there's the katakuchi, which is more like a bowl with a spout. Because it's wide open, there's much more contact with oxygen, which can help mellow some more intense sakes that are a bit rough around the edges. However, you don't want to be serving a kun-shu in it, as the aroma will practically disappear in such a vessel. Think of it like drinking a beer out of the maligned but still ubiquitous shaker pint glass. You don't get the best drinking experience in that case, as much of the aroma dissipates quickly with such a wide-mouthed container. But if it's a so-shu or a jun-shu with a very muted aroma, the katakuchi is an effective decanter.

If you can't keep all of these different items straight, don't worry about it. It's not likely that most places you'll drink sake will have a huge variety of drinking and serving vessels. It's more important to remember the traditional etiquette that goes along with serving and drinking sake. First of all, always pour for the other person. Keep an eye on their glass or choko. If it's close to empty, top it off. Don't let it get to the point where they're reaching for the bottle.

If you're pouring from the bottle, make sure you do so with both hands, with one hand on the neck and the other cradling the body. You'll always want the label side up. When you're done pouring, wipe the neck with a linen napkin or a serving towel and put the cap back on. During the pour, you should lay the cap closure side up to keep the germy table from infecting the side that will be in contact with the bottle's liquid contents. If you can put it down on a towel (still closure side up), even better.

Also, it's both bad form and hazardous to leave the metal break-away tamper-evident band dangling from the cap. Just twist it a few times and it will snap off. Don't try to pull it or you could end up in the ER awaiting stitches (I sound like my mother).

Lastly, when it comes time for you to drink, sip it; don't shoot it. It always makes me cringe when I see someone drink sake like it's bargain-basement tequila. For one thing, there's no way you're going to taste it if you just gulp it down that fast. Nihonshu is relatively expensive, so why waste your money if you're going to bypass most of the flavor experience? Plus, shots are just plain silly because this is a beverage that's usually around 15 percent ABV—or, in the case of gen-shu, around 18 or 19 percent—an alcohol content that's not radically different from that of wine. And you wouldn't do a shot of Chardonnay, would you?

SAKE COCKTAILS

I was debating with myself for months over whether I'd include a section on sake cocktails in this book. I've never been a fan of mixing sake in anything, but I had a mini-epiphany: you could say the same exact thing about just about any alcohol that's ever been shaken or stirred behind the bar. I like to drink whiskey neat, sometimes with a single cube or with a drop of water. But I also like a good Old-Fashioned, Manhattan, or Sazerac. There are some whiskeys I'll never mix (single-barrel bourbons, single-malt Scotches, and Japanese whiskies that are 12 years and older, — a Redbreast 15 single pot still Irish whiskey, for instance), but there are probably more that I'm perfectly happy having in a mixed drink. So why can't the same be said for sake? I'm not going to mess around with a junmai daiginjo with a seimaibuai of 35 percent or a robust and creamy yamahai, but surely there are plenty of futsu-shus, honjozos, junmais, and ginjos that might add a little something special to a cocktail.

There's also the matter of liquor licenses—some restaurants and bars can't sell spirits, which means that sake is a viable option along with beer, wine, and cider. And those sorts of venues often are looking to jazz up their menus with lower-ABV cocktails with non-spirit bases. Sake can replace many clear liquors in classic drinks, depending on the flavor profile you seek. Vodka enjoys the reputation as the most mixable spirit, but let's be honest, its neutrality makes it pretty boring. Vodka may have sake beat on alcohol strength, but sake definitely has a leg up in character. Here are a few sake cocktails that are worth trying, if you're into that sort of thing.

SAKE-TINI

Yes, this is a byproduct of the martini and cosmopolitan boom of the late 1990s and early 2000s (thanks, *Sex and the City*). As a clear alcohol, sake is an easy addition to—or even a replacement for—vodka and gin. The main appeal here is that drinking one of these won't leave you down for the count like most martinis will. A sake-tini has 100 percent of the classy, aesthetic elements—007's favorite stemmed glass, colorful garnishes—with about 50 percent of the booze, depending on the recipe.

SAKE-FORWARD SAKE-TINI

- ▶→ 2.5 ounces sake (go with a ginjo or junmai ginjo; the drier the better)
- ▶→ 1.5 ounces of gin (or vodka, but gin is the more authentic ingredient)
- ▶→ Ice
- ▶→ Japanese pickles (tsukemono) for garnish. If available, use takuan (pickled daikon), umeboshi (pickled plums), shibazuke (shiso-enhanced pickled eggplant and cucumbers), or any combination of the three.

Pour the sake and gin into a shaker with the ice and stir (forget what James Bond says; stirring is the way to go). Strain into a martini glass and garnish with the shibazuke (preferably combined on a toothpick). You may ask, Where's the vermouth? The sake serves that purpose—especially when its aromas are on the fruitier and more floral side of things.

GIN-FORWARD SAKE-TINI

Okay, if you're into something a bit boozier and botanical-centric, swap the volumes on the sake and gin.

- 2.5 ounces London dry gin
- 1.5 ounces sake
- Tsukemono for garnish
- Ice

Repeat all of the same steps above.

GIN-LESS, VODKA-LESS SAKE-TINI

If you want to eschew the Western spirits entirely, a possible alternative is to treat the sake as if it were gin or vodka and umeshu (Japanese plum liqueur) as if it were vermouth. This is definitely the more moderate of the sake-tinis, as umeshu typically has a lower ABV than most sake. You're going to want to go as dry as possible with the sake, something with an SMV of at least +5, to offset the sweetness of the umeshu.

- 3 ounces sake
- 1 ounce umeshu
- Tsukemono for garnish

Repeat the steps from before. If the sake and umeshu are kept chilled before mixing, you won't need the ice. (Why risk diluting it when you don't have to?)

SAKE GREEN-TEA COCKTAILS

*Tea punches and other cocktails are all the rage
these days, so why not give the concept some
distinctly Japanese flavor by combining it with one
of Japan's most famous non-alcoholic beverages?
If you can, stay away from major brand-name,
mass-market teas (especially if there's anything
related to the stars or heavens in the name) and
find some good artisan varieties. If you want a
more intense tea taste, go with matcha.*

- 4 ounces brewed green tea
- 1 lime
- Ice
- 4 ounces dry sake (junmai or junmai ginjo)
- Mint sprig

Brew green tea (loose leaf with an infuser, if possible). Juice the lime. Pour the tea in a cocktail shaker
filled with ice when done. Let sit for at least 1 minute
and then pour in the sake and lime juice. Give a
few good shakes and then strain into a tumbler or
Collins glass. Rub the mint around the rim of the
glass and then use it as a garnish.

125

SAKE BREEZE

*This one comes courtesy of Ronnie Prince, manager
at Chicago's Murasaki Sake Lounge. It's a little
more complex than the cocktails I've mentioned
thus far.*

- 1.5 ounces citron vodka
- 1 ounce Sudachi chuhai (a citrus-flavored distilled alcohol)
- ¾ ounces Seitokua junmai ginjo (or comparable
junmai ginjo)
- ¼ ounce elderflower liqueur
- ½ ounce Blue Curaçao
- Slice of lemon for garnish

Pour the vodka, chuhai, sake, and elderflower in a
shaker and shake for 15 seconds. Pour into a chilled
cocktail glass. Layer the Curaçao with the back of a
spoon. Adorn with a slice of lemon.

SAKE BOMB

Just kidding.

That's all I'm going to say about sake cocktails because I don't want
to actively encourage such behavior. Okay, I'm joking (kind of), but it
really is best to drink the stuff on its own. If your guests are skittish
about trying sake, though, this is perhaps the best way to ease them
into it while steering clear of the bomb zone.

CHAPTER 13

SAKE IN NORTH AMERICA, PART I: THE BARS AND RESTAURANTS

ear not! You don't have to be in Japan to have an epic nihonshu-sipping experience. There are plenty of watering holes and eateries in every time zone in the United States that offer drinkable tours of sake's birthplace without ever leaving American soil. Whether you're boning up for an impending trip, tiding yourself over until the next one, or just want to have a good drink close to home, here are a few of my favorite places to imbibe.

127

NEW YORK CITY

New York's vast sake-drinking scene kind of hides in plain sight. You really have to be in the know because the most prominent izakayas don't exactly telegraph their existence. But they don't need to because they're always jammed.

⇉ Sake Bar Decibel, 240 East 9th Street
You can't utter "Manhattan" and "sake" in the same sentence without visiting Sake Bar Decibel. What I love most about Decibel is that since its founding in the East Village in 1993, it's managed to retain its neighborhood's punk-rock roots as just about everything else changed around it. There's ceiling-to-floor graffiti on every wall of the damp, dank, subterranean pub, and they don't take reservations. You just grab a seat in a small holding area adjacent to the restrooms and wait behind a velvet rope for one of the servers to guide you to an open spot (waiting time rarely exceeds 30 minutes on a busy night). The drinks list features about 100 different sakes, as well as perhaps a

dozen shochu brands, and a handful of mostly Japanese whiskies and beers (not to mention cocktails). Pair them with any of a number of izakaya-style small plates such as the kurobuta sausages—they resemble cocktail weenies, but they're far more flavorful. If you're more adventurous, I recommend the dried stingray fin, one of Decibel's most famous dishes.

»→ Sakagura

Sometimes you're looking for a more refined experience than the loud and gritty vibe Decibel offers. For that, you'll need to take a trip uptown to Midtown East (where a lot of Japanese expats, extended-stay business travelers, and diplomats live). Sakagura is a stone's throw from Grand Central Terminal, so it's incredibly easy to access via commuter train or subway. Actually finding the upscale izakaya is another story, though, as it's located in the basement of an office building, down an industrial-looking stairwell (you have to pass the building's security desk to get there). But honestly, you really wouldn't want it any other way if you're seeking an authentically Japanese experience. It's common for such venues to be tucked away in structures housing various white-collar places of business in Tokyo and other densely populated regions of the country. The best part about it is that there's very little evidence of the outside world once you're inside. The windowless seclusion enables you to be truly transported and to immerse yourself in more than 200 nihonshu selections and an extensive menu of Japanese tapas. Go ahead and order five or six of them (if you're not dining alone).

Sakagura also happens to be the rare izakaya that's open for lunch, but reservations are a must. A typical lunch taster includes an array of fresh sashimi (including salmon, fluke, tuna, and amberjack), crispy shrimp and green pepper tempura, grilled black cod, sliced grilled beef in teriyaki sauce, and dessert.

»→ Juban

Juban fits in quite well among the galleries in Chelsea, as it's quite the artsy destination. Take some time to marvel at the paintings on the wall if you don't believe me. I like to think of this spot at Tenth Avenue and 22nd Street as the izakaya's izakaya. It's always a fun place to go with a group—as long as you remember to book a large enough table in the back ahead of time. You can get all of the classic munchables—tofu agedashi, assorted Japanese pickles, and gyoza—as well as some rather addictive skewers (sausage with ponzu mustard). There's also sushi and sashimi, but there's so much other good stuff you forget raw fish is even an option. About two dozen by-the-glass sake selections await you as do a number of bottle-only selections. Save the latter for the aforementioned group outings—you're not going to want to limit yourself to just one. A trip to Juban would also be a good time to try some shochu, as the menu regularly features 12 or 13 fine examples of the spirit (usually involving rice, sweet potato, barley, and buckwheat bases), all available by the glass.

⏩ Sakamai

This Lower East Side gastropub features the best of everything Japanese: food, whisky, shochu, and, of course, sake. Its staff typically includes a highly knowledgeable sake sommelier; Jamie Graves, a friend and invaluable resource for this book, once held that role. On a non-sake note, Sakamai was the regular home of Shochu'sday, American shochu expert Stephen Lyman's Tuesday evening showcase for the spirit before he started rotating New York venues and taking the two-hour pouring extravaganza on the road.

The candle-lit, exposed brick vibe makes it an optimal venue if you're trying to impress a date with your knowledge of fine Japanese alcohol. The staff mixologist stirs up plenty of craft cocktails, both bespoke and classic.

129

PORTLAND, OREGON

Portland can do no wrong as far as adult beverages are concerned, and that certainly applies to its nihonshu scene. The number of people who call the small city home is about a tenth that of the Big Apple, but Portland boasts a huge concentration of world-class sake-sipping sites. It very likely can boast the highest per-capita consumption of the beverage in North America.

"It's an unofficial claim," says Marcus Pakiser, vice president of the sake category for the Estates Group, a division of major wine and spirits distributor Young's Market Co. "It's a small category that no one's really keeping track of, but it's based on population and the amount of sake we sell here."

Pakiser has been building the Oregon sake market for more than two decades, first at brewery SakéOne and then on the distribution side. Sometime between 2005 and 2018, he says, Young's Market sake sales grew from about $400,000 a year to $3.5 million. "And that's just our company," he points out.

Sure, Portlanders have gravitated toward sake faster than consumers in other cities because, well, it's Portland, and that's just what they do.

"Most of America knows that this is a really good market for restaurants—it's a very chef-driven culture," Pakiser notes. "And chefs are very adventurous."

Perhaps a bigger causal element in sake's Portland market penetration is that it's one of the rare cities in the United States home to bars and restaurants offering single-serve sake.

"We do it like the Japanese do it," Pakiser says. "We use 1.8 liter and 720-milliliter—sometimes it's a 750-milliliter—and we'll pour it and sell it by the glass."

It's much more common for menus to sell by the 300-milliliter or 720-milliliter bottle and maybe offer one or two "house" sakes by the glass (one hot and one cold). Problem is, if you're bottle-only, you're not driving trial. A 300-milliliter of a junmai ginjo or daiginjo will run anywhere from $15 to $30 at a restaurant, while a 720-milliliter of the same sake will usually be upward of $60. Few will commit to that.

Meanwhile, look at your average wine list and you'll see far more selections available by the glass. And if properly refrigerated, sake actually lasts longer than wine does after the bottle is opened.

"[Restaurants] think sake goes bad in seven days, but they've really never tested that," Pakiser adds. "I tell people three months—and a lot of sake can go longer than that. If you're going to sell wine by the glass versus just by the bottle, why not sell sake by the glass? You're going to expose more people to it and then they're going to buy more of it."

I decided to experience the city's epic sake scene one rainy January when I built a multiday pub crawl that eschewed beer, wine, spirits, cider, and any other drink for which the City of Roses is famous. My goal was twofold: never drink the same sake twice, and never drink a sake I'd tried previously. As it turns out, those were simple objectives to achieve.

➻ Yama Sushi & Sake Bar

Everything you need to know about this Pearl District destination is in the name—and it doesn't disappoint on either front. Yama scores immediate points for offering wild-caught salmon sashimi and nigiri as a regular and reasonably priced menu item. Anago (sea eel) and Hamachi belly (fatty yellowtail) are also permanent offerings, ticking two more boxes in the "plus" column.

Another positive of Yama Sushi & Sake Bar is the visual splendor. The wall behind the sushi chefs is replete with nihonshu bottles—more than enough eye candy to keep me occupied while I dined alone. And they're not just ornamental. At any given moment, there are around 40 selections, and not just the unimaginative old-standbys that dominate the menus at lesser establishments. I try to keep a list of all of the sakes that I've consumed over time, which I consult whenever I'm drinking at a new place. I was shocked by the number of items on Yama's drinks menu that weren't already on my list (there might have been four or five that I had tried before). And the beverage director seems to keep things as fresh as possible, as the range of offerings on the website differed somewhat from the printed menu.

I must admit, however, that there were a couple of mildly disappointing aspects, mostly related to the manner in which the sake is served. The mouth of the glass was a bit too wide for my taste, which made it difficult to nose any nuances that may have been present. The server presented the glass inside a lacquer masu, but when she poured it, she stopped short of the ritual overflow. That was probably by design, since the glass was wide enough to fill most of the area inside the box. There really wasn't anywhere for the spillover to go. A solid argument as any in favor of slim, narrow-mouthed glassware.

⇥ Zilla Sake

You'll find no drinking vessel-related issue at Zilla Sake House in northeast Portland's Alberta Arts District. You get the narrow-mouthed glass and the overflow into the masu below. The only problem is that Zilla offers nearly one-hundred different sakes by the glass, and you'll never be able to narrow it down to just one or three for the evening's sipping escapades. Fortunately, there are several thoughtfully designed flights to give you smaller pours of three at a time. I tend to avoid flights at most places that offer them because there are at least one or two in each that I've either had before or just aren't that exciting. Not at Zilla, whose flights provide a little something for everyone. And it's not obvious groupings like "fruity and floral" or "sweet" versus "dry." I gravitated to the "Yasei Mai" flight, which features a trio of nihonshu made with three different varieties of heirloom rice (Watari Bune, Mikinishiki, and Omachi, on that particular grouping). That's some next-level stuff right there.

"I do a monthly flight that rotates depending on whatever I'm feeling," says Kate Koo, Zilla's co-owner, who possesses certification from world-renowned sake expert John Gauntner's Sake Professional Course and the Wine & Spirit Education Trust (WSET) Level 3 Award in Sake. "I'll do a kimoto flight, or a nama flight when it's nama season. Our goal is to not only help introduce people to the world of sake, but also help people who've already got their foot in the door."

If you just want to go drink and maybe have a few light snacks at Zilla, you can grab a stool in the bar area (a separate room) and do so without feeling hounded about ordering a full-on sushi dinner. If you prefer the latter, you're lucky enough to be in one of the most authentic sushi restaurants in the city, with ever-changing seasonal catches. Personally, I'm a fan of sitting at the bar and chatting with the passionate, knowledgeable sake experts pouring the drinks. I once ordered a kimoto junmai from the menu, which inspired the bartender to pour one for himself and join me because he hadn't had it in a while. That's the sort of place it is.

Zilla's happy hour will usually feature inexpensive pours of Momokawa from nearby SakéOne brewery. You can also indulge in the specially priced maki and onigiri of the day, which showcases a rotating roster of fish.

Just make sure you make at least one trip to the bathroom while you're there. Hundreds of sake labels wallpaper the restrooms.

"We called in a couple of favors and asked for different labels, maybe some things that were a little unique, the Japanese versions of labels of bottles that we have here," Koo recalls.

I warned my wife that one day she's going to come home from work and find our bathrooms transformed with similar décor.

⇥ Yataimura Maru

I could easily include all of the Portland izakayas within the Shigezo group, the owner of Yataimura Maru, but I'm picking this one because it reminds me of the sort of vibrant and noisy, no-nonsense spots you'd find in a major Japanese city. It also represents a textbook reason for why you should never judge a book by its cover. The cover in question was the front vestibule, where I noticed a printout pasted to the door identifying several key Japanese phrases spelled phonetically. It struck me as a bit kitschy, and I worried Maru would be a Disney-fied version of an izakaya. But my fears were allayed the moment I stepped inside.

The constant sizzle of the fryer amidst rising steam and dangling lanterns really nails the vibe for me. It doesn't hurt that the air is always redolent with takoyaki, chicken karaage, and yakitori. The sake list offers something for every taste, and there's always an addendum with several limited-time-only selections, so every time you visit you're likely to find something new. Plus, the servers seem ready to answer any question knowledgeably. That's really all I ask for out of life.

⇥ Masu

I've already told you how in love I am with the traditional, cedar, 180-milliliter box after which this downtown Portland spot is named, so you'll understand why I had to include it. (And it's right around the corner from Powell's Books, and you know you were going there anyway.)

The entrance is not readily apparent, as the street-level door opens on to a staircase that leads to the second-story restaurant. Once upstairs, you'll discover rustic horizontal wood paneling (more of a classic Japanese look than a '70s rec room vibe) and a mural of Mount Fuji. But my favorite part of the décor dangles from the ceiling: a mobile made out of the namesake wooden boxes.

Food-wise, Masu leans more toward traditional sushi bar than izakaya (though there's plenty of yakitori to be had), but it does have a bit of fun with its raw fish dishes. If you like salmon, the menu features a flight of five different kinds of those upstream swimmers. Before your appetizers and main selections show up, you might be surprised to discover a wooden masu full of edamame arrive at your table (they're really selling the box concept).

Even if you weren't the least bit hungry, you could still make a day of working your way through the sake menu—which not only includes detailed tasting notes, but seimaibuai info as well! And you can probably guess the manner and the vessel in which it's served.

The bonus is that Masu is open for lunch, so if you're game to start sipping early (and who wouldn't be?), avoid the evening crowds and luxuriate while the muted blue carpeting absorbs any surrounding noise there may be.

⇥ Bamboo Sushi

For a time, this was the only sushi place I chose to frequent any time I visited Portland. And while the food is amazing, it's the nihonshu that has etched Bamboo forever in my heart. I rarely remember the first

133

bar I tried a particular drink, but I know that Bamboo is where I was introduced to Kanbara Bride of the Fox junmai ginjo—which likely will have a permanent place on my top 20 sake list for decades to come. Another key selling point for Bamboo: the food menu is 100 percent sustainable, meaning you'll find no nigiri, sashimi, or rolls featuring any overfished species. So if you're looking for Hamachi, you're out of luck. There are multiple Bamboo locations throughout the city; I usually gravitate toward the Southeast Portland site on SE 28th Avenue.

↠ Chandelier

Chandelier is one of the newer arrivals in Portland's sake scene and it's definitely found a place on my short list. It's super-tiny, which I love because some of the best places in Japan to drink barely fit a half-dozen people at a time (Chandelier has room for at least a few more than that). The menu typically has around twenty-five different sake options, divided by flavor. But the best part is the decor. Sure, there's the namesake light fixture, but what keeps me going back is the David-Lynchian design. There's a red curtain and black-and-white zig-zagged linoleum floor, emulating the Red Room at the entrance of the Black Lodge in "Twin Peaks."

LOS ANGELES

When all else fails, head to Little Tokyo. It's usually a safe bet for finding good sake (and sake importer Mutual Trading Co., which also houses the Sake School of America—where I took the sake sommelier class and exam—just happens to be located nearby).

↠ Ohjah Lounge

I usually wouldn't recommend a hotel bar for sake, but Ohjah Lounge is one of the few exceptions. Located on the second floor of the Miyako Hotel, it's the standalone lounge connected to the restaurant and sushi bar Tamon-Ten (damn fine in its own right, and you can order from the same sake list to sip with your dinner). But if you're not in the mood for a meal or just want light apps and snacks (takoyaki, Japanese pickles, cold tofu, etc.) with your beverages, Ohjah's open from 5:30 p.m. to 2 a.m. seven nights a week. There's a full bar, but you're really here for the sake. You'll find an extensive selection of premium yet affordable options, and there will always be a special bottle or two that the bartender can't wait to tell you about. Oh, and there's karaoke (and if you're like me, that's the game-changer right there!).

SAN FRANCISCO

Historically, San Francisco has enjoyed significant Japanese influences, and those are quite apparent in the city's izakayas.

▸▸ Nara

Nara offers a modern, gastropub take on the classic Japanese watering hole, with hardwood walls, communal tables, and attractive minimalism. Aside from a few plants, the main décor is a collection of nihonshu isshobin (those 1.8-liter bottles) of different varieties and colors. You'll find at least two dozen offerings, encompassing junmai, ginjo, daiginjo, nigori, honjozo, and sparkling sake, as well as a handful of shochu options, yuzu, and plum liqueurs and even the cloudy Korean sake cousin makgeolli.

WASHINGTON, DC

In recent years, the nation's capital has really come into its own as a full-fledged food-and-drink destination (many credit having a foodie president, Barack Obama, with helping to turn that around). And as District gastronomy improved, so did the sake-drinking scene.

135

▸▸ Daikaya

Despite its location across the street from the Capital One Arena (the venue formerly known as the Verizon Center), Daikaya feels like a neighborhood joint. It's actually two restaurants in one—the ground floor houses a casual ramen bar, while upstairs offers a more trendy izakaya experience. One of the go-to staples is cod roe spaghetti, which, in addition to the namesake fish eggs and noodles, features nori (dried seaweed), chervil, and cream sauce. Another is humba, braised pork belly with Okinawan black sugar and salted black beans. I'm a big fan of its well-curated sake list, usually featuring nearly 40 selections. And it's not just because the choices are stellar but because it's organized by style. There's a section for ginjo/junmai ginjo, daiginjo/junmai daiginjo, futsu-shu/honjozo/junmai, and the harder-to-classify "others."

▸▸ Sushi Capitol

Located in the Capitol Hill neighborhood, Sushi Capitol offers, hands-down, the best sushi in the District—and the greater DC area, for that matter. The sake menu is no slouch either. This is not the type of place you'd want to frequent every week, as it can be a bit pricey, but it's great for celebrating birthdays, anniversaries, and any other special occasions during which you don't mind getting a bit spendy. The best part about the drinks menu is that it offers both 720-ml and 300-ml bottles; the latter comes in handy when you don't want to commit to just one for the entire evening. My recommendation: sit at the sushi bar, order the omakase, get a succession of three 300-milliliters of different styles (or double that if you're a party of four), and Uber home.

⟫→ Sushi Taro

A very close second to Sushi Capitol among sushi restaurants in Washington, DC, Taro, I would argue, slightly edges out Capitol when it comes to sake selections. However, I must offer a word to the wise from personal experience: read the menu carefully. I once saw $55 listed next to a bottle of Oze no Yukidoke junmai daiginjo and thought it sounded reasonable enough to spend on a 720-milliliter for two people. When we got the bill for the meal, I was a little confused by the fact that it was $300, pre-tip. Then I realized that I'd neglected to see the number next to that "$55." The $55 option was the 8-ounce carafe, just a touch more than a single serving. The 720-milliliter was a much heftier $150. Live and learn (it was damn good sake, though).

136

CHICAGO

Chicago's a drinking town, through and through, and its modern cocktail scene is one of the most iconic in the entire world. But don't let that overshadow the Windy City's quietly stellar sake scene.

⟫→ Izakaya Mita

In addition to at least three dozen carefully selected sake offerings of all stripes (you can pretty much taste your way through all of the major premium styles and then some), Izakaya Mita also sports one of the Midwest's most extensive shochu lists (and the Japanese whisky menu isn't too shabby either).

⟫→ Booze Box

Upstairs you can get your raw fish fix at Sushi Dokku. Downstairs it's all izakaya—the more upscale variety, anyway—in the best low-lit manner imaginable. The West Loop pub offers about a dozen sake selections, but the list is more about quality than quantity. There's nary a dud among them. If you're curious or just plain indecisive, Booze Box offers sake tasting flights—bartender's choice, so you know it's going to be curated with care. About three-quarters of the selections are available by the glass and by the 720-milliliter bottle (a couple are only available in 300-milliliter bottles and the house hot sake only in an

8-ounce carafe), and the food menu includes some izakaya staples like takoyaki (fried balls with chunks of octopus in them) and okonomiyaki (savory Japanese pancakes with eggs, shredded cabbage, and other delightful morsels), as well as some maki rolls, nigiri, and sashimi courtesy of the sushi bar upstairs.

And it's not just the run-of-the-mill stuff. There often will be fresh and rare catches that the server is more than happy to expound upon (not to mention upsell, because it's usually pretty pricey stuff). The waitstaff is quite attentive as well; on my visits they've been able to spot near-empty sake glasses from across the room and take it upon themselves to top them off, provided there's anything left in the bottle we'd just bought. Booze Box isn't exactly a cheap night out (unless you want to stick to beer and gyoza), but it's nice for those treat-yourself occasions.

▶→ Murasaki Sake Lounge

There's been a Japanese-style bar on East Ontario Street in Chicago's near North Side/River East area since the very early '90s, but it hasn't always been Murasaki. For the first two decades-plus of its existence, it went by the name Café Shino and could best be described as a vaguely divey, carpeted, Japanese snack bar with white walls and a piano that was straight out of the '80s. It was a popular spot with Japanese businesspeople who were either passing through the Windy City or sticking around for an extended stay. But times change, as do owners.

When bartender Jun Takanarita bought the place in 2014, he seized the opportunity to give the bar a much-needed makeover. The space became much darker, much more intimate (candlelit, of course), and exponentially more modern, with a marble bar and comfy seats. The vibe is still 100 percent Tokyo, but more the twenty-first-century version of the Japanese capital city. Takanarita traded the piano for a turntable, and guest DJs now spin every Saturday and the first Friday of each month. Japanese movies play on the flat screens (with the sound off) to enhance the sense of place.

Murasaki typically carries about 30 different sake selections, an equal number of Japanese whiskies, roughly 20 shochu offerings, and 8 Japan-brewed beers. And it happens to be the only place in downtown Chicago where you'll find private karaoke rooms. You'd have to venture to the city's Chinatown or Koreatown to find the others.

"Our servers are very well versed in Japanese sake, as well as whisky and shochu," says manager Ronnie Prince. "When the guests come in,

137

they can get an education on what's a trendy sake or what's the best sake to consume with our dishes—[we're] kind of introducing both Japanese and American people to fully understand the process that goes into creating these drinks."

Murasaki has sake cocktails on the menu, but they're less a specialty and more of a jumping-off point for newbies.

"Guests can try [sake] as a cocktail first so they can get the hang of it and see that there are different ways to enjoy it," Prince explains. "But we try to focus on the base. If they like the cocktail, we try to let them taste the sake independently so they understand the full flavor of it."

And despite its adherence to Japanese nightlife traditions, Takanarita infused it with a bit of hometown pride: expect to find, alongside the Asian imports, local craft spirits brands like Koval and Few on the shelves.

DENVER AND BOULDER

Even at a mile high up, you can still find some pretty stellar sake experiences if you know where to look.

⇥ Izakaya Den
Located adjacent to its sister eatery, Sushi Den, Izakaya Den is definitely the side you want to be on if you're more interested in drinking than inhaling raw fish. Izakaya Den offers close to three dozen sakes to choose from with a variety of staples most regular izakaya-goers would expect in such a venue: agedashi tofu, tempura, gyoza, and something called "Seafood Dynamite" that features baked baby octopus, scallops, rock shrimp, calamari, mussels, crabmeat, mushrooms, mayonnaise, and masago. If you're currently in Denver, what are you waiting for?

⇥ Amu
Do yourself a favor and drive the approximate 35 miles from Denver to Boulder, and enjoy what's likely the most authentic Japanese eating and drinking experience in the Rocky Mountain region. Make sure there aren't any holes in your socks because the moment you walk in, you'll be asked to remove your shoes.

CHAPTER 14

SAKE IN NORTH AMERICA
PART II: THE BREWERIES

F irst it was craft beer. Then it was craft spirits. And then, on a smaller scale, craft cider started to have its moment. But what we don't hear too much about, at least in mainstream media channels and within the US food and beverage world at large, is craft sake. I wouldn't necessarily call that unfair, considering how minuscule the movement is in America. It is, however, something worth keeping an eye on, as the number of American artisanal sake producers is growing slowly but steadily.

Sake production on US soil is not a new concept, as some larger Japanese producers set up American breweries decades ago. Technically, American sake production dates back to 1908 when the Honolulu Sake Brewery first opened to produce the Takara Masamune brand (of course, it didn't officially become an "American" brewery until Hawaii achieved statehood more than a half-century later). However, as far as modern brewing in the continental United States is concerned, it all began in earnest in 1979 when Ozeki, a Hyogo Prefecture company that dates back to 1711, commenced production at a Hollister, California, facility, about 100 miles southeast of San Francisco. You're likely to spy Ozeki's value-priced junmai on the shelves of liquor stores and supermarkets with even the most modest sake selections.

Three years after Ozeki brewed its first American batch, Takara Shuzo founded Takara Sake USA in Berkeley, California, less than a decade before Honolulu Sake Brewery closed for good. Then, in 1983, Takara Sake USA launched the Sho Chiku Bai line, which has since expanded from the original junmai to include nigori, junmai ginjo, junmai daiginjo, and organic nama offerings. Not long after Takara

139

opened its Berkeley production facility, Fushimi-based Gekkeikan's Folsom, California, plant went on line in 1989. The launch of Gekkeikan Sake (USA) Inc. has enabled the company to grab about a 25 percent share of the total US sake market, most of which is thanks to its flagship junmai, Gekkeikan Traditional. Other US-produced Gekkeikan bottlings include Black & Gold, a blend of two sakes—one with a 60 percent seimaibuai and the other, 70 percent—to add an extra level of complexity; Haiku tokubetsu junmai, the low-temperature fermented Silver; and its nama, Gekkeikan Draft. You're likely to find few honjozos here as US laws make it cost-prohibitive—adding even just a small volume of distiller's alcohol to the mash means the beverage now must be taxed as a spirit.

If I were to borrow some beer terminology, Gekkeikan and Takara would be akin to the "macro" brewers, despite the fact that their output is a relative drop in a bucket next to the likes of beer's biggest players. However, those original US sake makers helped develop Americans' taste for the Japanese beverage and pave the way for the new wave of Stateside craft producers.

If there's ever been a greater sign that it's strike-while-the-iron's-hot time for sake in America, it was famed nihonshu brewery Asahi Shuzo's announcement in 2017—35 years after Takara set up shop in Berkeley, and 30 years after Gekkeikan USA opened its doors in Folsom—that it would be building its first US facility outside of Poughkeepsie, New York, to expand production of its iconic Dassai line. By the way, "iconic" is not overselling it. When Japanese prime minister Shinzo Abe visited the White House in 2015, President Obama toasted with Dassai 23—its 23 percent seimaibuai offering. Word is, Dassai is going to be making a daiginjo that retails for $15 per 720-milliliter bottle— unheard of for a daiginjo-grade sake.

Dassai's arrival likely serves as the vote of confidence in the category that prospective entrepreneurs need to further spur small sake startup activity. And for the most part, I am really liking what I'm seeing so far. Some of the newer players are crafting some pretty world-class stuff, as good as many premium offerings produced in Japan. Others, not so much (but you didn't hear it from me).

Here's a small peek at what's happening in the United States.

—— SAKÉONE ——

Forest Grove, Illinois

It's only fitting to kick off this section with the one sake brewery that's arguably done the most to advance the case for sake making in America. And, ironically, brewing wasn't even a part of SakéOne's original business plan when it first opened in 1992. Its initial mission was to import brands from Japan's Momokawa Brewery, and after five years of doing that, it decided to start making its own. SakéOne launched the Momokawa brand as a tribute to its Japanese partner. The Momokawa line features several junmai-ginjo offerings, each with its own flavor nuances: Momokawa Ruby is lightly sweet, with tropical fruit and cherry notes; Momokawa Diamond is more on the medium-dry side of the SMV scale, with some melon and orchard fruit on the nose; Momokawa Silver is more full-on dry and crisp with a touch of citrus; and Momokawa Pearl is a sweet, rich, and creamy nigori with banana and dessert confection notes. There's also a Momokawa Organic Junmai and Momokawa Organic Nigori. Beyond the Momokawa trademark, SakéOne offers the "G" line of genshu sake, packaged in a stunning, trendy-nightspot-ready opaque black bottle, and the Moonstone range of fruit-infused junmai ginjo.

141

If you find yourself in Portland, Oregon, a 20-minute drive outside the city will get you to SakéOne's headquarters, which hosts tours of the production facility and tastings in its gift shop. When the weather's nice (which, admittedly, is an infrequent occurrence in Oregon), you can nose and sip glasses of Momokawa on the outdoor deck.

I like to think of SakéOne as the baseline for all of the American craft sake producers that have come (and gone) since. I visited the Oregon brewery a couple of times in 2013 and 2014, right before the newest wave of US craft sake producers got up and running. I returned in early 2018 after having tried the products from many of the younger operations, just to see how my impressions of both SakéOne and American sake in general had evolved, if at all.

My biggest takeaway is that there's really something to be said for consistent quality. I feared I would succumb to that beer geek's notion that anything that's been around a while and has reached a certain size—SakéOne is small by beer standards but huge by non-Japanese sake standards—is yesterday's news. But I'm happy to report that wasn't the case. In fact, SakéOne is the brewery that younger operations should aspire to be—not necessarily in flavor profile, but in batch-to-batch quality. I'm on board with the newbies' desire to establish an

American sake identity, stylistically speaking, but I also believe they should appreciate that the classics are classics for a reason.

On my 2018 trip to SakéOne, CEO Steve Vuylsteke let me in on a little secret: Japanese importers have expressed interest in bringing the Oregonian products to their country. If you're an American brewer that's able to export its sake to Japan, you're obviously doing something right.

BEN'S TUNE UP
Asheville, North Carolina

If the name hasn't already tipped you off, Ben's is a bit of an odd duck. Located at the site of a former auto garage, there is some logic to the moniker of this brewpub/sake brewery/beer garden. Ben's brews its own sake from American rice and has a few different varieties on tap—yes, on tap—at any given time. One of those might be fruit-infused, another might be a nigori, and the other might be a nama genshu. Where the flavors and aromas of SakeOne's offerings hew fairly close to tradition stylistically, those for Ben's Tune Up can be all over the map with a love-it-or-hate-it intensity. There are plenty of Asian and Polynesian-inspired bar bites—even a pupu platter—to provide a bit of balance for those loud and boisterous sake flavors.

Until recently, Ben's was one of two sake breweries in Asheville. The other, Blue Kudzu Sake Co., shuttered in 2015. The owners cited a long permitting process that kept the products from reaching the market and generating revenue among the reasons for shutting down. It also didn't have the multiple revenue streams that Ben's enjoys as a brewpub, bar, and restaurant. It wouldn't survive on sake alone.

PROPER SAKE CO.
Nashville, Tennessee

Asheville isn't the only Southern city with a storied adult beverage culture that has welcomed a sake brewery within its borders. Its rhyming cousin in neighboring Tennessee has become as well known for its craft beer (Yazoo Brewing, for one), artisanal spirits (shout-out to Corsair Distillery), and thriving cocktail scene (Patterson House, Bastion, and Urban Cowboy Public House are among my ever-growing list of faves). So it was only a matter of time before sake joined the fun. And it did so in the form of Proper Sake Company, which set up shop next to the Nashville location of urban oenological phenomenon City Winery—as if to say, in a not-so-subtle way, "Move over." Byron Stithem, Proper's founder and brewmaster, applied his culinary expertise to Music City's first sake production facility, having built a distinguished career at such respected establishments as Dinner Lab, Husk Nashville, and the Clover Club. He opened his brewery's doors at the tail end of 2016 and has since released a trio of entries to the nascent American craft sake category. First is the Diplomat, a junmai brewed with No. 9 strain of sake yeast that's meant to be a crowd pleaser. Hence, its name: the brewery says it offers a diplomatic balance of complexity and sweetness. Then there's the Diplomat Unfiltered, a nigori version of its flagship junmai. Things start to get experimental with Grand Parlay, a junmai brewed with a saison yeast commonly used to make fortified wines. The reason behind its name? It serves as a "parlay" between Eastern and Western cultures, marrying a bit of European wine-making tradition (giving Grand Parlay a little bit of white wine-esque astringency) with the signature alcoholic beverage of Japan.

Another intriguing hybrid was Smoky Mountain, aged in a quinoa whisky barrel (supplied by Nashville's Corsair Distillery) and cold-smoked with pecan wood. Stithem admits that it was a parlay of a different sort, as it's meant to draw in folks who aren't too keen on trying sake. Its woody, whisky-like elements are a way of saying, "Here, this might taste familiar to you."

Proper's Super Gold Sparkling has some fine, subtle, naturally occurring bubbles and might even appeal to those who say they're not into sparkling sake (the menu says it's "like an umami-driven Prosecco," which is true to an extent, but I think that description undersells it just a bit).

All of Proper's creations are nama, or at least nama-adjacent. Some of the distributed bottles sit for a moment in a 147-degree-Fahrenheit bath. And that's really to mitigate spoilage in the event that the retail account doesn't handle it properly.

"It's not technically Japanese nama, but it's not technically pasteurized," Stithem explains. "It's heated a little bit to make sure it's shelf stable. I tried to go full nama to start, but had a few problems. You can't really tell where people are going to store it."

If the flavor ends up just a little bit off, the consumer is more likely to judge the brewer than the seller or server. Stithem also eschews dilution; everything is genshu, usually around 18 percent ABV. He says the fuller flavor of undiluted sake is a bit more attractive to American palates.

"When I first started drinking sake, it was nama genshu," he says. "To be able to share that with people has really been my dream. Even in Japan, where sake sales kind of teeter down but go up in the rest of the world, you're starting to see the youth culture interested in fuller flavored drinks like whisky and beer. I think this style will resonate with them as well."

Proper boasts an apt tasting room in the recently revitalized Pie Town neighborhood on Nashville's south side. It's a space that rewards the curious. If you were to stroll through the post-industrial district, you'd never know that behind one of the many unremarkable metal delivery bay doors lies a mini oasis of Japanese wooden minimalism. Stithem says he and his business partner were able to take their time on the meticulously appointed décor because it took a while for the brewery permits to come through. What else were they going to do, let the space collect dust and get all cobwebby?

The tasting room houses a bar (with towering stools; I almost hurt myself when I was getting off of one because I forgot how long the drop was!), some communal tables, and a couple of more intimate nooks for smaller groups and dates (I once witnessed a large group gathered to play Dungeons & Dragons). Proper gets a surprising amount of walk-in traffic, mostly locals who are pleasantly stunned that a sake brewery even exists in their city. It helps that some craft beer breweries and distilleries have been opening up in the neighborhood, which is turning into Nashville's unofficial artisanal beverage district.

I decided to play "fly on the wall" (well, "fly perched on a very tall stool") and just listen to some of the questions and conversations that were orbiting the bar. One of my favorites:

Man: "Do you have hot sake or is it just cold?"

Stithem: "We serve everything chilled."

Man: "I do love sake; it's just usually whatever they have hot at the restaurant."

A sake brewer's job as educator is never done (full disclosure: Stithem sneaked me a sip of Diplomat—the non-nigori, non-sparkling one—heated to 123 degrees Fahrenheit, which I'll detail in my tasting notes).

"It's been really refreshing to see how many people come in with no prior knowledge of sake, except what they had at sushi restaurants—scorching-hot rubbing alcohol, essentially," Stithem says. "It's been cool to open up their eyes to the wide variety that sake can offer. Especially in the South, it's not the style of sake that was readily available. Previously that education point, that push, wasn't really there."

On occasion Stithem has even opened the eyes of what he calls sake "purists."

"Generally, when you get the purists, they're looking for that ginjo/daiginjo style, so I really can't appease them at the moment if that's the only thing they're looking for," he reveals. "But I've gotten some good feedback from them."

The best part of tasting at the source, aside from ambiance, is the fact that you get to sample the stuff on draft—a rarity anywhere in the United States. If you like what you taste and want to take it home, Proper packages its sakes in opaque, 8-ounce screw-cap bottles that you easily could mistake for cocktail bitters or shrub containers. It's a great trial size for you and a guest.

While you're sipping at the bar or gazing away from your role-playing companions for a moment, you might be intrigued by what lies behind a translucent plastic curtain. Purple and magenta lights beam down over what appears to be some sort of botany lab. Proper shares the space with an indoor microgreens nursery that supplies fresh veggies to local restaurants.

Stithem's own culinary connections have enabled Proper to partner with Nashville restaurateurs beyond just selling them his beverage. He's gotten the foodie scene hooked on koji.

"This is another way to hopefully share the wealth and science of sake," Stithem notes. "Koji is one of the most amazing instant fermentation starters around. Most of the restaurants in town that have any sort of progressive style are using some kind of koji base for something."

Trendy, industrial-chic eatery Rolf & Daughters, for instance, bakes a bread with the spores, and its diners are crazy for it. (And I understand why; Stithem had a little stash of it behind the bar, which he was kind enough to let me taste.)

It wasn't the easiest journey getting the koji to Proper, let alone to any of its restaurant partners. As you can probably guess, there aren't too many labs cultivating the spores for commercial sale, given the size (or lack thereof) of the American sake-making industry. Stithem had to procure his cultures from a supplier in Japan—a family operation that has been producing koji spores for more than two centuries. But it wasn't as simple as taking a trip and making a purchase.

"Basically, I had to go over there and plead my case for them to sell to me," he remembers. "They're very protective of their product, especially when it leaves the country. I essentially had to go hang out with the son of this family for three days. He took us to all of these great places in Kyoto and Osaka, and at the end of the trip he was like, 'I will sell you koji spores.'"

Of course, those are microorganisms we're talking about. They don't take up too much space. Now imagine you wanted the same kind of water that's made Japanese sake famous. There's no sensible way, logistically, to get huge quantities of Miyamizu from Nada. But that doesn't mean you can't try to replicate it in Nashville, which is exactly what Proper Sake has done. The brewery uses water that has been augmented to contain a mineral composition mimicking that of the Hyogo Prefecture's legendary source. If you ever visit Proper, be sure to ask for a glass of water. Hydration is very important!

—— MOTO-I ——
Minneapolis, Minnesota

Since 2008, Minneapolis has had its very own sake brewpub. Moto-i produces junmai, junmai genshu, junmai daiginjo, junmai nama, and junmai nigori products; and its sister company, Minnesota Rice and Milling, polishes its own rice as well as rice for other sake brewers who don't have their own milling machinery. The founding of Minnesota Rice and Milling was a smart move on the part of Moto-i, as sake rice polishing isn't too common a practice in the United States, and it removes one of the barriers to entry for those entrepreneurs who've been kicking around the idea of launching a sake brewery. This way, they can have Minnesota Rice and Milling polish their rice to their precise specifications without having to throw down the cash for a vertical polisher right away.

Moto-i boasts a pretty extensive food menu as well, serving a variety of Japanese and pan-Asian dishes like season-specific sashimi, yaki udon, and bulgogi.

—— BROOKLYN KURA ——
Brooklyn, New York

It's generally a pretty good indicator of a trend's potential longevity when it arrives in Brooklyn. The borough now rivals Portland, Oregon's status as the craft beverage capital of North America. In early 2018, the multiple-award-winning beer breweries and distilleries that hang their hats in Brooklyn welcomed a sake maker to their side of the East River. Brooklyn Kura, which proudly dubs its products "American craft sake," opened its brewery and tap room in the Sunset Park neighborhood's Industry City, an arts, retail, and dining complex that's also home to Industry City Distillery.

"Originally this was going to be distillers' row," said partner and brewer Brandon Doughan as he showed me around the production facility during its first month of operation. "We were like, 'Let's not tell them that we're not distillers!'"

147

Doughan met co-founder Brian Polen at a mutual friend's wedding in Tokyo in 2013. After many glasses of sake, their dream was born.

"Brandon left research science, I left corporate America to do this," said Polen as he filled my glass with the brewery's flagship junmai ginjo nama genshu, #14.

Doughan honed his sake-making craft at a pair of breweries in Japan as well as at SakeOne in Oregon.

After much experimentation and fine-tuning, the Brooklyn Kura team would lock in the recipe for #14 as well as the formula for Blue Door, its more robust junmai, and Lake Suwa, its full-bodied, fruit-forward option. The Outer-Borough-chic, post-industrial tasting room also offers blends and alternate iterations of the main products, such as a version of Lake Suwa that's a shiboritate—a type of sake that goes right from pressing to packaging without the usual six-month resting period. The brewery also lets visitors taste an orizake, a style that retains some of the (usually filtered-out) sake lees that settle to the bottom of the aging tank.

"We can present sake at different points in its life cycle," Polen notes. "We also serve little tastes from the vat and let people taste it next to the finished product."

Not only does Brooklyn Kura enjoy the distinction of being the borough's first sake brewery, it's also the first in all of New York City. You'd think such a diverse, cosmopolitan, foodie-friendly market would be easy to crack, but Polen reveals that it's a lot harder to navigate than you'd think.

"If there's any market in the United States that has expectations about sake," he says, "it's this one."

—— BLUE CURRENT BREWERY ——
Kittery, Maine

Maine is a beer state as much as Brooklyn is a beer borough. And the former's status, most would agree, preceded that of the latter. Now the same can be said for sake. Blue Current Brewery, located just across the New Hampshire border (a stone's throw from Portsmouth) in Kittery, had a few years' head start on Brooklyn Kura, and now Blue Current's junmai ginjo is starting to pop up around New England. It's also starting to get international attention; it took a gold medal at the 2016 London Sake Challenge (a pretty prestigious event, considering the fact that Harrods hosts it). The brewery uses California-grown Koshihikari rice (polished at Minnesota Rice and Milling, Moto-i's sister company) and spring water sourced near the coast in its own state (you know, the place where Poland Spring comes from).

Founder and toji Dan Ford was working in the financial technology world until the 2008 banking crisis. After that, he decided to turn his passion for sake into a business. Ford received both the Certified Sake Professional and Advanced Sake Professional certifications from the Sake Education Council in Tokyo. He initially launched Blue Current in his garage and stayed there until 2015, when he moved the business to a full-fledged standalone facility that includes its own kojimuro.

Ford sees a significant opportunity to woo drinkers of white wine, especially since he sees a kinship between fruity ginjo-grade sake and lovers of pinot grigio. He's had some success in doing so, as he's managed to get Blue Current's sake in a number of non-Asian accounts in Maine and other parts of New England. And it's no mean feat convincing potential customers that the beverage is as much of an ideal pairing for a burger as it is for yakitori or sushi. Ford admits it's a constant struggle. "It's very difficult to sell," he told me as I gazed around his brewery—a building that used to house an auto repair garage. "When we make a sale, I almost have to educate my customer—especially if someone doesn't carry it in their store. I get questions like, 'What is sake, how's it made, what's it made from?' and then I'm like, 'Here, taste it.'"

Sometimes he'll run through that process in reverse, not even telling potential retail accounts and consumers what they're actually tasting at first: "'Just taste it, tell me if you like it,' I'll say. 'Oh, I love that,' they'll say. 'Oh, that's sake? What? No, sake's hot.' It's always like this. People have a preconceived notion about everything."

Blue Current still gets considerable support from Asian venues even though they may initially get a little sticker shock.

"I'm handcrafted, not machine-made, so my price point is a lot higher," Ford says. "They get addicted to this box sake—it's cooking sake, really. For a small bottle, we're almost three times that at retail."

—— DOVETAIL SAKE ——
Waltham, Massachusetts

Maine isn't the only part of New England boasting its own sake brewery. Waltham, Massachusetts, in the greater Boston area, is home to Dovetail Sake, co-founded in 2014 by culinary school graduate and Certified Sake Professional Daniel Krupp and craft beer industry veteran Todd Bellomy. The brewery produces the fruity and light Nakahama Junmai, with a seimaibuai of 60 percent and an SMV of +5. It's named for Nakahama Manjiro, a nineteenth-century sailor credited as the first Japanese person to live in the United States. He was rescued by a New England-based ship and ultimately settled in Fairhaven, Massachusetts. Manjiro became the first Japanese citizen to command a trans-Pacific voyage, serve as an officer on an American vessel, and ride a steam train. So it's only fitting that a Massachusetts brewery would name its inaugural sake after such a pioneering individual with roots in Japan and a home in New England.

Dovetail's nigori, known as Omori, also has a bit of history behind its moniker. American zoologist Edward Sylvester Morse (from New England, of course) was on a research trip to Japan in 1877 when he observed what came to be known as the Omori shell mounds (he was passing through Omori train station when he first spotted them). Hidden within those white shell mounds were artifacts dating back to an era known for its pottery and the introduction of rice farming. Not only do the shells and the nigori share a milky-white hue, both are full of surprises. It's a bit on the sweet side with an SMV of -10.

It's an exciting time for sake in the Northeast, and I can't wait to see what else happens as more sake makers in the region master their craft.

⸺ GAIJIN 24886 ⸺
Denver, Colorado

When they launched Gaijin 24886 back in 2014, business partners Marc Hughes and Keith Kemp really wanted to give drinkers a sense of their hometown terroir.

Hughes had been a home brewer (it is Colorado, after all), but he was really looking to steer his hobby toward fermentation traditions that weren't all that common within the home brewing community. He dabbled in wine and cider, but didn't find much satisfaction there, as he wasn't passionate about either of those beverages. He then turned his attention to sake and was initially surprised by the dearth of how-to information that existed for the beverage in the public sphere. Eventually he found an old FTP site where a brewing enthusiast had archived an old sake recipe that legendary beer writer Fred Eckhardt (who passed away in 2015) had published. (Though more famous for his beer work, Eckhardt was also an evangelist for American sake. He published a regular newsletter focused on the beverage and was the author of *A Complete Guide to American Sake, Sake Breweries and Homebrewed Sake*.) Hughes then got his hands on some books translated from Japanese that fortified his amateur sake-making endeavors, and he and Kemp went pro sometime after.

Few of us traditionally think Rocky Mountains when we think of sake, but the Gaijin guys are hoping to change that with the most Colorado thing to happen to the category perhaps ever. Among Gaijin 24886's offerings is the dry and fruity Queen City of the Plains, named after home city Denver's nickname in the late nineteenth and early twentieth centuries. And if you weren't already clear on where this stuff is made, there's El Colorado, which sports a reddish-amber hue as a nod to origin of the state's name (Colorado means "colored red").

One thing Gaijin 24886 has been proving is that Coors Light doesn't have a claim on Rocky Mountain spring water. For the Golden, Colorado, mega-brewery, fetishizing the water used to make its brand is really just marketing (especially when it tries to convince the public that it makes the coldest beer around—which is essentially meaningless). But with Gaijin, the water is really its soul. It's not unlike some of the liquid flowing in Kyoto, which has made that city's sake world famous.

Gaijin 24886 takes the pristine-water concept even further for some of its experimental, smaller-batch efforts. For the special release, the brewing team hiked up a mountain to collect snow that would become the liquid base.

Beyond such complex innovations, the brewery's main focus has been developing products that are more approachable for American consumers. "We try to tweak the flavor profiles to be more effective for the American palate," Hughes says. "Some of the ginjos and daiginjos have a lot of delicateness, but they don't necessarily go with [the American] diet, which tends to be fat heavy, sugar heavy, and salt heavy. We get much more of that banana flavor in our sake, where it'll be a much lighter fruit flavor on the Japanese palate."

151

Like Blue Current, Gaijin 24886 makes it its mission to dispel the myth that sake belongs only in Asian venues.

"I think there's still this focus that sake has to only be with Japanese food, with sushi or ramen, and only at a Japanese restaurant," Hughes says. "I think it's changing, sort of slowly."

He likens that perception to the same sort of attitudes that above-premium tequila marketers have had to combat.

"There was a lot of really bad tequila in the world," Hughes notes. "And [marketing] was focused only on that college, 'Taco Tuesday' sort of thing, where people do a lot of shots of tequila and get blackout drunk."

Similarly, mainstream American consumers think of sake as that stuff they heat up in sushi bars because it's too vile to drink cold. Unfortunately, for decades, there hadn't been enough higher-end sake being imported to or produced in the United States and making its way into Japanese restaurants. So it's taking a while to change those perceptions.

The good news is that tequila is finally having its moment both as a sipping, connoisseur's beverage and, along with its cousin mezcal, a darling of the craft bartending scene. So maybe sake isn't that far behind (though I'm ambivalent about sake as a cocktail ingredient). At the very least, it's encouraging that Gaijin 24886 seems to have the Mountain Time Zone covered on that front.

— TEXAS SAKE COMPANY —
Austin, Texas

I admit, I chuckle a little bit every time I utter the words "Texas Sake Company" out loud. But it's totally legit. Founded in 2011 in Austin (where else?), Texas Sake offers two flagship products, Junmai Nigori and Junmai proper, and has been known to experiment a bit with such products as Peachy Keen ginger peach sparkling sake and Oak, a wood-aged version of its junmai.

Texas Sake Company didn't really hit its stride until 2014 when the original owner sold it to Adam Blumenshein and Tim Klatt, the owners of the nearby Strange Land Brewery (the craft beer kind). They hired toji Jeff Bell, who revamped the products and processes. Previously, Texas Sake had used a variety of brown rice grown in the Lone Star State that wasn't in the Japonica family and, therefore, not exactly suitable for sake production. The Texas-grown rice made the original products overly starchy, and it also made the sake sour too quickly, according to Texas Sake spokesman and sales manager Trevor Wight (Wight adds to the brewery's cross-category pedigree, having worked in wine and beer production, as well as distribution, prior to joining the company). The new proprietors began sourcing the California-grown Calrose variety (polished to 70 percent in the Golden State before it makes its way to Texas), which has become the varietal of choice within the burgeoning American craft sake industry (though, in an effort to keep things as close to home as possible, Texas Sake Co. has been looking into eventually sourcing rice grown across the Texas border in Arkansas—more on that in the sidebox).

Despite the fact that the new Texas Sake team has dramatically improved production and ingredient quality control, they aren't necessarily interested in mimicking flavor profiles established in Japan. On the contrary, the brewery really wants to lead the charge in developing a truly American style of sake, using only US-grown and sourced raw materials. Japanese shuzo kotekimai might be nothing short of legendary, but Texas Sake's goal is to put some Stateside breeds on the sake-making map (and, judging by Wight's email signature, they're also trying to make "Kanpai, y'all!" a thing). In late 2017, Texas Sake Company signed a distribution deal with mega wine-and-spirits wholesaler Republic National Distributing Company, which helped the brand expand its reach well beyond the Austin market. That, in turn, necessitated a scale-up in production in 2018. In other words, expect to be hearing a great deal about the little Texan shuzo in the years to come, especially as the American craft sake scene matures beyond its infancy.

—— SEQUOIA SAKE ——
San Francisco, California

San Francisco's Sequoia Sake takes pride in California's bounty, using Sacramento Valley Calrose rice and water from Yosemite National Park. Founders Noriko Kamei, Warren Pfahl, and Jake Myrick are also fairly fastidious about the microbes, selecting kojikin whose ancestry is linked specifically to sake making for hundreds of years.

And, as it happens, Calrose has deep ancestral ties to Japan as well.

153

"I lived in Japan for 10 years," Myrick notes, "and I got to know hundreds of breweries and people in the industry. I sent our rice and water to labs [in Japan] for them to analyze before we got started. When you're sourcing koji, you want to make sure the koji is matched to rice."

Although the initial koji strain turned out to be a negative match, researchers discovered that Calrose, first grown in the Sacramento Valley in the early twentieth century, actually was a descendant of the Watari Bune breed—one of the first grains ever used for sake making. Watari Bune may not be a variety we hear all that much about these days, but we certainly are quite familiar with its famous offspring: Yamada Nishiki.

Sequoia's core line includes a junmai-nama, genshu, and nigori for distribution at the brewery and select regional retailers.

"Ninety-nine percent of the sake [from Japan] that you here get in the United States is pasteurized because it's mandated by the government," Myrick says. "To get nama to the United States is very difficult."

Sequoia's core unpasteurized line makes it a bit easier for Bay Area folks to get the full namazake experience. "If you have it with a salad, it'll taste like melon," Myrick says, "and if you have it with seafood, it'll taste like a dry Riesling."

The company also markets its Coastal line, which features bottle-pasteurized ginjo, genshu, and nigori, designed to travel better beyond Sequoia's immediate region.

The production team brews sake every 45 days to ensure optimal freshness.

Though initially launched as a hyper-local operation serving only its immediate region, 2017 and 2018 marked a period of great expansion for Sequoia, during which it hired its first staff members beyond the founding three and doubled its production capacity. It has since made its way south into the lucrative Los Angeles market.

"We're trying to coin the same concept that California wine did in the '70s," Myrick reveals. "We're American craft sake; we're not trying to be Japanese craft sake."

Myrick is optimistic about the bona fide American craft sake scene that's emerging throughout North America. "Consumers really like the whole tradition of craft and there's a great story for sake," he says. "It's easy to understand, similar in that way to beer and wine, but it's a cleaner beverage. There are no tannins, no glutens, no sulfites. I'm not going to say it's 'healthier' because we can't talk about health."

Sequoia may be a young company—it opened in 2014 and launched its first product in 2015—but it's never too early to think about the next generation. When I spoke to Myrick in early 2018, his daughter, Olivia, was in the middle of a five-year internship at Akebono Shuzo in Aizu, Fukushima, and completing the Fukushima Sake Academy's three-year sake master's program. "If she comes back and wants to take up the reins," Myrick says, "I'm all for that."

— SETTING SUN SAKE BREWING CO. —
Miramar, California

Sake brewing has a presence in Southern California as well. Craft beer industry vets Josh Hembree and Keldon Warwick Premuda—who both previously worked for the San Diego area's most famous craft producer, Stone Brewing—opened Setting Sun in Miramar, California, in 2015. Like Sequoia, it too is using rice grown in the Sacramento Valley, which is the closest thing Americans are going to get to premium shuzo kotekimai this side of the Pacific. Setting Sun has been pretty innovative with its barrel program, aging sake in Chardonnay and Cabernet barrels (and even dry hopping some of them because old habits die hard for beer people)—the first American sake brewery to take such a beery approach to sake making.

But having a craft beer background doesn't just foster innovation. Hembree, who brought more than a decade of brewing experience to his role—two at Stone (his partner was there for ten) and eight at a series of US and international operations—says that it serves as an ideal foundation for those looking to transition to the sake world. It gave him a hands-on education in brewing science, especially as it relates to the way yeast behaves (it's also helped brewers who've wanted to leave beer behind for cider or mead).

—— ARTISAN SAKE MAKER ——
Vancouver, British Columbia, Canada

155

I'm going to squeeze in one Canadian brewery. Sake expert and lecturer Masa Shiroki opened Artisan Sake Maker on Vancouver, BC, Canada's Granville Island in 2007 and has since produced a line of junmai sake—including a nama, nama genshu, nama nigori, and sparkling sake—under the brand name Osake. For the first several years of its existence, Artisan imported all of its rice from Japan and the United States, but the company always hoped it would be able to make a product from 100 percent Canadian ingredients. After years of research and trial and error, the brewery was able to develop a local rice breed suitable for sake production. In 2013, it released, for the first time, batches that included only British Columbia–grown and sourced components. Nothing goes to waste at Artisan, as it also sells its sake kasu (lees) for use as a culinary ingredient (the brewery's website includes several recipes for those interested in experimenting a bit with the leftover solid components from the nihonshu-making process).

A Bit About American Sake Rice

Most US sake brewers have been sourcing their rice from the Sacramento Valley, which grows a variety known as Calrose—technically the umbrella name of a group of different cultivars descended from traditional Japanese sakamai (if you want to get a bit technical, each of those specific cultivars has a number attached to it: 201, 204, 205, 208, etc.). It mimics Japan-grown sake rice in most ways, with one significant distinction: it lacks shinpaku, that starchy heart. The starch is still there; it's just a bit more dispersed throughout the kernel. Milling machines are still able to polish away all of the unwanted layers of fat and protein—they just take some of the starch with it, as it's freely floating in space among those other less desirable (for sake) nutrients.

Calrose used to be the only game in town, but now there's honest-to-goodness Yamada Nishiki growing on US soil—in Arkansas of all places. And that could be a real game-changer in the American craft sake world.

Chris Isbell of Isbell Farms is a true pioneer. He was the first to successfully grow Japanese-style rice in the early 1990s. At that time it wasn't a sake variety but a sushi one: koshihikari. Prior to that, conventional wisdom insisted that it could not be done outside of Japan. His efforts attracted a great deal of attention; Japanese tourists and businesspersons visit Isbell Farms regularly. He's even sold some of his grain to Japan.

Not long after Isbell proved the naysayers wrong about koshihikari, he started to explore whether he could grow, as he calls it, "a more lucrative variety." He read up on Yamada Nishiki, figured out how to plant and cultivate it, and accomplished yet another task they said couldn't be done.

"I planted it just to look at it as a grower, how it worked agronomically," Isbell recalls. "I planted a row of it, looked at it, made notes on it, and then put it in the freezer."

Then several years later, he got a call from Takara Sake USA.

"They wanted to know if we had Yamada Nishiki, so we started to grow for them to practice and tested it for five years or so," Isbell says. "They were very excited, and it's been ramping up since then." Now he's regularly fielding calls from start-up sake breweries looking to get in on the Yamadanishiki action.

"I've talked to two just this past week," he reveals, noting that he'd just finished a call with one of them minutes before we spoke.

Isbell has formed a symbiotic relationship with Minnesota Rice and Milling. He will do some preliminary milling—to 90 percent seimaibuai, just to turn the brown rice to white—and then ship it off to Minnesota's Blake Richardson.

Isbell certainly welcomes the potential to expand his business as the American craft sake scene grows.

"We're geared up, we're ready for it," he says. "Our farm has 3,000 acres. If we had to grow that much, we could. We're only growing about 100 acres now."

· CHAPTER 15 ·
TASTING NOTES:
SAKE, AMERICAN-STYLE

This latest series features only sake brewed on US soil—including the newer crop of craft producers as well as the American operations of Japan-based brewers.

SHO CHIKU BAI ORGANIC NAMA

Brewer: Takara Sake USA
Location: Berkeley, California
SMV: +5

Sho Chiku Bai Organic Nama was produced in Berkeley, California, at the brewery that serves as the US branch of Kyoto-based Takara Shuzo. Fruit tends to dominate, but it's more sour apple than anything else—I'd even go so far as to say apple cider vinegar. There's even a slight hint of salinity. Nama sakes can have surprising elements (they're alive, after all), and Sho Chiku Bai Organic Nama never fails to surprise.

MOMOKAWA SILVER

Brewer: SakéOne
Location: Oregon
SMV: +7

When someone relatively new to sake asks for a recommendation of a good, dry, crisp sake, my suggestion is almost always Momokawa Silver. SakéOne demonstrates what American-produced sake could (and should) be. The junmai ginjo holds its own against world-class examples of its grade produced across the Pacific. I also don't want people to get sticker shock when I recommend some of my favorite dry and crisp imported junmai ginjos, which will usually run at least $10 to $15 more per bottle than the Oregon-brewed Momokawa. Fruit aromas familiar to sake drinkers—melon, pear, and their ilk—and a bit of earthiness make it quite the pleasing experience, either with a variety of foods or all by its lonesome.

MOMOKAWA #701 NAMA NAMA

Brewer: SakéOne
Location: Oregon
SMV: +7

To get a full sense of how foregoing pasteurization and dilution impact the character of sake, taste Momokawa #701 Nama Nama—the number is the yeast strain—alongside Momokawa Silver. This one is the nama genshu version of Silver. In addition to the melon and orchard fruitiness of the regular junmai ginjo, #701 also boasts an expressive spiciness and a cranked up, rocky minerality. There's even a very subtle yet detectable fizziness, thanks to a little carbonation left over from the fermentation process. "Lively" and "jazzy" don't even begin to describe it.

G FIFTY

Brewer: SakéOne
Location: Oregon
SMV: 0

G (stands for "genshu") is SakéOne's line of undiluted sake packaged in opaque black bottles. The Fifty is the seimaibuai, which technically makes it daiginjo grade, but it's marketed as a ginjo (SakeOne eventually may introduce a daiginjo, but not until it's good and ready). There's a slight grapey quality here, tempered by a bit of pear, maybe apricot, and even a touch of saline minerality.

160

BLUE CURRENT JUNMAI GINJO

Brewer: Blue Current Brewery
Location: Maine
SMV: 0

The core product of Maine's first sake brewery would be the perfect choice to punctuate a meal. It has enough sweetness to give most dessert wines a run for their money as well as a creamy texture and fruity, confectionary notes not unlike marshmallow Peeps or banana pudding

THE DIPLOMAT

Brewer: Proper Sake Co.
Location: Nashville, Tennessee

The flagship junmai from Music City's Proper Sake has quite a lot of end-of-meal vibes, especially if your idea of dessert involves a lot of vanilla and cream, and possibly a cheese plate. There's also a certain tartness to it that evokes Key lime pie, as well as a wild and dry cider from Spain's Asturias region or Basque Country. There's no need to be diplomatic here—it's spot-on. And that's just when it's cold. I tried some heated to 123 degrees Fahrenheit, and it reminded me of a warm banana cream pie.

161

THE DIPLOMAT UNFILTERED

Brewer: Proper Sake Co.
Location: Nashville, Tennessee

"Unfiltered" is a bit misleading here. This is a nigori, and a nigori lover's nigori at that. It showcases many of the same flavor and aroma elements, but it cranks up the tartness even more. There's a surprisingly salty finish as well. Proper Sake Co. founder Byron Stithem told me that's probably due to the fact that the Diplomat develops some savory elements as it ages.

SUPER GOLD SPARKLING SAKE

Brewer: Proper Sake Co.
Location: Nashville, Tennessee

Finally, a sparkling sake to which I can offer a full-throated endorsement! It's essentially a bubbly interpretation of the Diplomat with a bit more grape and apple character. However, there's still a goodly bit of umami and a saline finish similar to that on the nigori version. The bubbles are much more understated than those in most sparkling offerings—which is probably why I'm such a fan of this one. If you told me the slight fizziness was just residual from the fermentation process, I would totally believe you. It's just that subtle.

SMOKY MOUNTAIN

Brewer: Proper Sake Co.
Location: Nashville, Tennessee

"This is my olive branch to people who don't want to drink sake," says Proper's Byron Stithem. Think of it as the nexus where sake meets whiskey. It's also where two of Nashville's great craft producers intersect: Proper and Corsair Distillery, which provide the Quinoa Whiskey barrels in which Smoky ages.

If you like woody flavors, this one's for you. That of course means all of the vanilla you could possibly want, as well as some hearty, nutty, toasted grain expressions. It also has some of that same salty character that emerges when Proper products mature.

GRAND PARLAY

Brewer: Proper Sake Co.
Location: Nashville, Tennessee

Grand Parlay is named so because, as Proper says, it represents a parlay between Eastern and Western cultures: sake and beer. Proper brews it with saison yeast, which gives it some of the same wild and earthy qualities you'd expect from a farmhouse-style beer. There's also a grapefruit tartness as well as some bready elements that remind me a bit of freshly baked pumpernickel.

#14

Brewer: Brooklyn Kura
Location: Brooklyn, New York

If you want to get a crash course in junmai ginjo nama genshu, I can't think of a better place to start than with Brooklyn Kura's #14. It's ticks all of the boxes in the fruit and floral realm, with notes of pineapple, violet, and even a bit of licorice shining through.

BLUE DOOR

Brewer: Brooklyn Kura
Location: Brooklyn, New York

Things get a little chewier and a bit more expressive with Brooklyn Kura's junmai. Craft beer drinkers who like the banana-ish elements of a typical hefeweizen are going to really want to open the Blue Door.

LAKE SUWA

Brewer: Brooklyn Kura
Location: Brooklyn, New York

Melon is the MVP of the flavor notes swimming in Lake Suwa. There's also a little marshmallow sweetness tempered by some sharp, crisp acidity.

MOTO-I JUNMAI GINJO NAMA ZUME

Brewer: Moto-i
Location: Minneapolis, Minnesota

The junmai ginjo from one of the most prominent new producers of American sake has a texture that's almost chewy—which is amazing—and flavor and aroma notes that see-saw between sweet cooked rice and tropical fruit. Yep, there's that pineapple again.

DOVETAIL NAKAHAMA JUNMAI

Brewer: Dovetail Sake
Location: Waltham, Massachusetts

The New England brewer uses 100 percent Yamada Nishiki in Nakahama Junmai, creating an engaging sake with a clean, fruity aroma and a dry kick. See if you can spot notes of strawberry and cherry. I did, but I might just be crazy. Dovetail used private yeast strains that it says no one outside of Japan has used for sake.

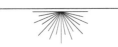

CHAPTER 18

SOME FINAL FOODIE MUSINGS

Aside from the upcoming epilogue (which, I'll admit, I wrote before I wrote this chapter), I was struggling to find a way to bring in this Japanese-inspired journey for a soft landing. And then someone shared an article that ran on the VinePair website in January 2018. It fit into the "You're doing it wrong" subgenre of lifestyle/food/drink journalism.

I've reached the point where I'm able to tune out such clickbait, but when the subject is sake, it naturally catches my attention. The gist of the piece was that you should never drink sake with sushi. Ever. The headline was one of the most grating of those I've encountered on the subject: "Americans Love Pairing Sake with Sushi, but It's a Big Mistake."

To be fair, the notion is somewhat rooted in tradition. There's a Japanese saying that advises people to "not have rice with rice."

But there are a few aspects that get lost in translation when Western journalists and restaurants try to espouse a zero-tolerance attitude toward nigiri and nihonshu pairings. For one thing, it's kind of an archaic philosophy. You'll mostly find older consumers of Japanese cuisine abiding by it. The younger ones kind of march to the beat of their own drummers.

The idea of pairing any alcohol with food is relatively new in Japan, in the same way that pairing anything but wine—namely, beer—with food is still a burgeoning concept in the United States.

165

The entire concept of "diversity" in sake didn't really exist in Japan before the 1970s. Breweries were just starting to cut their teeth on ginjo and daiginjo—and even then it was mostly to impress their peers and wine competitions.

"These very high-end styles were considered too esoteric and expensive to the general public and weren't widely available at the time," says New York sommelier Jamie Graves, who manages the sake portfolio at importer Skurnik Wines & Spirits.

Futsu-shu was pretty much the only option (and even four-plus decades later it still commands around three-quarters of the market).

"As the economy developed," Graves adds, "people were a little more willing to pay for something unique and different like that."

Times change and customs change. There's been a recurring theme at most of the traditional Japanese sake breweries I've visited: they've designed many of their flagship products to complement meals. And in many of those meals, sushi is involved. For the most part, diners in Japan continue to avoid drinking sake with dishes that include a heavy rice or other carb component. However, they tend to ignore the custom when they're eating sushi (sushi typically isn't what one would call a "heavy" meal). It's more about not wanting to feel too full from a carb-laden dish than concern over any sort of rice-on-rice taboo.

The author of the VinePair article talks to the beverage director at an upscale New York Japanese eatery, who says he tries to steer restaurant guests away from sake and toward beverages like Riesling when they're ordering nigiri and maki rolls. At that moment I had to put down my laptop and go for a walk.

This rant is just a long way of saying that, no, sake with sushi is not "wrong"—but not every sake is necessarily "right." I hope that if you've read this far into the book, you now have some sense of the vast diversity of flavor and aroma profiles brewers can achieve with just rice, water, koji, and yeast (and sometimes a little added distilled alcohol). Within that broad range, there are more than few gustatory and olfactory components that match quite well with sushi. And then there are others that go well with oilier cuisines, such as Chinese takeout. You might even find a few that are an ideal match for pizza.

Graves says nihonshu, regardless of style, tends to have one particular quality that other alcoholic beverages lack. "Sake is very forgiving with food in a way that wine isn't," Graves points out. "The level of acidity in wine means that certain wines can really lock in with some foods and either hit a bull's-eye or completely fall apart. With sake it's rarely such a high-stakes endeavor."

True, it's far less common to hit that life-altering bull's-eye with a sake pairing, but it's also less likely that it's ever going to completely miss the mark.

There are certain flavors that are just not meant to come anywhere near wine. Fish roe—long a component of the Japanese diet—will completely clash with the acids in wine. "That's why vodka is a traditional pairing with caviar," Graves notes. Sake may not be nearly as neutral a beverage as vodka, but it more than holds its own against briny fish eggs.

Anything that's super salty can find a loyal friend in nihonshu. Ask anyone familiar with the *otsumami*—the Japanese tradition of eating small, usually intensely salty snacks during drinking sessions. *Otsumami* items typically include delicacies like shiokara (sea creatures such as squid in a viscous paste) or assorted pickles that have frequently accompanied glasses, carafes, and bottles of sake.

167

There are generally two opposing forces at work with any successful pairing: complement and contrast. Decide which one you're aiming for and go from there. Kate Koo of Portland's Zilla says crab dishes provide good opportunities to test these concepts. On the complementary front, one particular sake immediately comes to mind for her: Taiheikai Tokubetsu Junmai from Huchu Homare Shuzo. Taiheikai means "Pacific Ocean," so that should give you an idea of what it offers: a bit of brininess, like the sea, especially when warmed up. There are also fruity and floral elements that pair with the sweetness of the Dungeness crab typically on Zilla's menu.

For contrast, Koo says you should lean toward something a bit drier, cleaner, and earthier. You'd be hard-pressed not to find a selection that fits that bill, though there is one particular style you might want to steer clear of.

"I probably wouldn't go as funky as a yamahai," Koo says, "because crab is a delicate flavor."

It's not likely you're going to encounter crab on a daily basis. Cheese, on the other hand, probably would make more frequent appearances in your diet (provided that you're not lactose intolerant or vegan).

It took decades of hop-and-malt evangelism to get American drinkers to grasp that beer is a better companion for cheese than wine (and, to be honest, most are still resistant to or unaware of that assertion, as I continue to get more invitations to wine and cheese parties than I do to beer and cheese parties). The fat-scrubbing carbonation in beer helps reset the palate. Even though most sakes lack carbonation—with the obvious exception of the sparkling sort—many varieties actually go quite well with cheese.

An aged sake—a juku-shu, if you recall—often makes a suitable partner for a mature cheese, like a cave-aged cheddar, Parmesan, or a three-year-old gouda. Aromatic sakes with light and delicate flavors—kun-shu—may be an ideal accompaniment for mozzarella or goat cheese. The rich, full-flavored varieties—those of the jun-shu ilk—can hold their own against funkier, more pungent selections. Yamahai iterations are especially desirable since much of the wild complexity from the lactic acid asserts itself in those.

168

Having said that, some more subtle sake styles are equally at home with such aggressive cheeses. Graves points to the engaging contrast between the more assertive elements of blue cheese and the fruity character of some junmai ginjos and daiginjos. The fruit notes have a mellowing effect on the salty, earthy funk of a Gorgonzola or a Stilton.

You're probably accustomed to ordering wine with pasta, particularly those in the red sauce family, but sake is just as versatile in the Italian-American arena. Granted, you're not likely to find much in the way of nihonshu at your neighborhood trattoria, but you'd do well to stock up on some bottles for a quiet night at home with a bowl of macaroni. For no-nonsense sauces like spaghetti pomodoro, select a pairing that's subtle in flavor but robust in aroma (a less delicate kun-shu—something you're not going to destroy with tomatoes).

If you're going for the meatier sort of pasta dishes, along the lines of Bolognese or a carnivore's dream of a Sunday sauce (my Italian-American roots are showing), the big, bold reds you might drink with those can easily be replaced by richly flavored and textured sake, something in the jun-shu neighborhood. The same goes for a nice, juicy, fatty steak (cooked medium-rare, or we can't be friends). The richer, the "meatier," the funkier the sake, the better it is for a deeply marbled cut of beef. The good news with steak is that it's a lot easier to test that pairing outside of your home, as izakayas with extensive sake lists frequently have some sort of grilled beef on the menu (the Japanese do have some of the best beef in the world, after all).

I mentioned pizza earlier, and I know you're not going to let me off the hook until I suggest some sort of pairing for that. There are a couple of different roads you can take, depending on what kind of pizza it is. If it's a sauceless, four-cheese kind of deal, you can apply many of the same principles you would to the cheese itself, depending on how sharp, earthy, and stinky it is. A yamahai of some sort would be best; a yamahai genshu would be ideal.

If garlic and spices dominate, you want something that's going to be a bit of a palate cleanser, something that has a clean crispness to it. A dry-ish junmai or junmai ginjo should fit the bill, as long as the umami elements are a bit muted in it. The same should apply for tomato-sauced pies.

Marc Hughes, owner and brewer of Denver's Gaijin 24886, has quite a bit of experience in this area, as he spends a great deal of time promoting sake pairings with Western dishes. Pizza is a frequent companion at such events. And that runs the gamut from simple and understated margherita to the carnivore's paradise that is a meat lover's pie. If it's particularly meaty and fatty—with toppings like pepperoni, prosciutto, and/or sausage—Hughes recommends an offering with some pronounced fruity elements, such as his brewery's El Colorado red sake. "Sake will also pick up some of those subtle oregano flavors that won't usually come through," he tells me, "and it helps temper down some of the spices."

Side note: If you're anywhere near Anchorage, Alaska (and I'd imagine only about two or three of you are because, you know, population), put some of these suggestions into action at Hearth Artisan Pizza, which usually carries eight or ten sake selections at any given time, along with the more traditional pizza-pairing beverages.

Hughes is also a fan of teaming his brands with the all-American hamburger—even when the patty is buried in toppings. Genshus of many grades can enhance the burger-eating experience, he says. "Undiluted sakes have that heavier, rounder mouthfeel to them," Hughes explains. "And those help, sort of, to keep all of that flavor in your mouth."

169

Hughes also gives a shout-out to what he considers one of the quintessential burger-friendly sakes: Kikusui Funaguchi, famous for the single-serve can in which it's packaged (check out my tasting notes on page 46). "It has a lot of full-bodied flavor and it coats really nicely and allows a lot of the flavors to sit in your mouth for a long time."

Since we're on the subject of American staples, we can't not talk about barbecue. Yes, beer and bourbon go great with it, but you're not always going to be in the mood for either of those. Jake Myrick, co-founder of San Francisco's Sequoia Sake, is a big fan of a genshu with the saucier items in the barbeque universe.

Now, let's say you want to move out of the realm of classic American comfort food and are in the mood for something a bit more daring, like a South Asian curry.

I've been to a number of Indian restaurants offering early-bird specials (okay, I'm old) that included a complimentary glass of red wine. I never could understand why that was a thing because I don't think there's a beverage that's a worse match for Indian—and spicy dishes in general—than wine. But it was quite a popular promotion.

I spent most of my life in New Jersey, where local governments were stingy with alcohol licenses, and there were more BYOB Indian restaurants than those that served any kind of booze. More often than not, people brought red wine to those establishments.

You probably know where I'm going with this. Sake—particularly a fruity ginjo or daiginjo—is a much better option to pair with spicy curry dishes.

"It's like having chutney with your curry," Graves suggests. "You still get all of the spice and flavor with the curry but with the fruity sensation."

If all else fails, just remember that light-flavored sakes pair well with equally light dishes. So something in jun-shu's zip code would be fine with a salad, particularly one with a light vinaigrette or a citrus-based dressing, rather than a heavy, creamy one.

On the flip side, sake of deep, robust flavor goes well with dishes of deep, robust flavor—think stew and chili, for instance. When you remember those simple quasi-rules, it's pretty simple to extrapolate from there.

But we're not done eating quite yet. I hope you've saved room for dessert, because sake has as much of a place at the end of a meal as it does at the beginning and the middle. In fact, not only are some varieties of nihonshu ideal accompaniments for tasty confections, many have actually become those confections.

When I visited Midorikawa Shuzo, president Shunji Odaira greeted me with green tea and something that resembled a slice of pound cake but was considerably more moist than those I'd previously encountered. As it turns out, sake was the key ingredient in the cake. And it wasn't just a gimmick; I could actually taste its presence, and it was quite delicious.

I left another brewery with a handful of individually wrapped sweet gelatin snacks made with the kasu left over from the production process. And while I was waiting for my train to arrive, I came across what I had thought was just an urban legend: sake-flavored Kit-Kats (the Japanese Kit-Kat game is strong—you also can find green tea, apple, strawberry, and a host of more esoteric flavors of the candy bar there).

A little later I stumbled upon some soft-serve sake-flavored ice cream, which at first sounded like a bit of a head-scratcher, but upon further research (forking over ¥400 for a small cone), I was very pleasantly surprised at how well it actually worked.

171

Sadly, you're not likely to encounter such sweets at your local greasy spoon or Baskin Robbins. So unless you're willing to try your hand at baking, chilling, or freezing your own (and the internet offers no shortage of recipes), your best chance of having the sake-dessert experience is to, well, have sake with dessert.

This is another area that's relatively new to the industry, which means that it fits in with the recurring "there are no hard-and-fast rules" theme.

Some have recommended pairing fruity sakes with fruity confections, but I'm not necessarily a fan of that fruit-on-fruit notion when it comes to dessert. I mean, you wouldn't put strawberry jam on fresh strawberries, would you? I'm more of a strawberries and cream guy, so if you're eating fresh strawberries, maybe try a junmai nigori to have a similar taste experience.

I'm also a fan of bananas and cream or bananas and chocolate. If you're having, say, a bowl of chocolate or vanilla pudding or ice cream, find a sake with very distinct banana-like aroma notes. The pairing can be something of a revelation. Just make sure it's not too dry. Normally I'm a fan of contrast, but when it comes to dessert, dryness can clash in a rather unpleasant way.

You almost can't go wrong with a sparkling sake with dessert, especially the fattier ones (is there any other sort?). Again, there's that scrubbing-bubbles notion.

Now I want to share a minor pet peeve of mine. You don't need to agree with me, but hear me out:

There's a common misconception, even among many in the sommelier community and the sake industry, that when compared with other fermented beverages, sake's flavor spectrum is quite narrow. That kind of thinking doesn't do anyone any favors. There's obviously a vast gulf between, say, an imperial stout and a Belgian-style witbier, and the stylistic contrast isn't necessarily that stark in the nihonshu category. But taste a yamahai junmai genshu next to a daiginjo and it's a night-and-day experience—even though they may look exactly the same. Like many of its producers, sake is a humble beverage. It doesn't broadcast its flavor diversity like beer and wine do, but once you spend some time with the beverage and get to know it a little better, you discover that there is so much that lies not too far below the surface. I'd even argue that flavor variety is much more extensive in sake than it is in wine. Aside from the obvious red-versus-white divide, how many non-connoisseurs—about 99 percent of all drinkers—can really tell you, right off the bat, the difference between a Cabernet or a Malbec? And sake's many nuances—like the complexities of a new human acquaintance—become increasingly apparent when explored over a meal.

I've said it before and I'll say it again: tasting is highly subjective. You know what tastes good to you, so never let anyone try to tell you what you should like, regardless of how much more experienced they are as sommeliers, chefs, or just plain eaters and drinkers. It's their job to listen to you and find the right match for your preferences. And you know what . . . if you want to eat a big heaping bowl of fried rice and wash it down with a carafe of sake, don't let them try to suggest a nice, dry Sauvignon Blanc.

CHAPTER 17
TASTING NOTES:
A FEW MORE FOR THE ROAD

I wouldn't want you to leave empty-handed (or empty-glassed). Check out these selections from all over Japan.

MINAKATA JUNMAI EXTRA DRY

Brewer: Sekai Itto
Prefecture: Wakayama
SMV: +15.5

In case the name and SMV weren't clear enough, this is the dry sake lover's sake. It sports a light, mildly complex aroma of melons and grains with a dryness that lingers for quite a while but never overstays its welcome. If you're looking for something to pair with tempura and other fried foods, you really can't go wrong with this one.

TENRYO KOSHU JUNMAI DAIGINJO "IMPERIAL LANDING"

Brewer: Tenzan Shuzo
Prefecture: Saga
SMV: +3.5

"Imperial Landing" lands in an area that's somewhat fruit-forward—banana in particular—with some licorice and faint starchy and nutty elements that manifest in a gentle touch of fried rice on the back end. There's a round richness in the mouthfeel of this aged sake, which makes it an adventure to pair with a wide range of meats.

MICHINOKU ONIKOROSHI HONJOZO

Brewer: Uchigasaki Shuzoten
Prefecture: Miyagi
SMV: +10

Honjozo is so often misunderstood, and Michinoku Onikoroshi is one of the best defenses against that misunderstanding. There's a goodly amount going on here, including some notes of cooked rice, hazelnuts, maybe some mushroomy earthiness, and a touch of banana. It's great chilled, but a little bit of heat brings out some of the umami elements.

BUNRAKU JUNMAI DAIGINJO

Brewer: Bunraku Sake Kura
Prefecture: Saitama
SMV: +2

Notes of prickly pear, kiwi, and pineapple waft up to the nose from this delicate junmai daiginjo that's incredibly smooth and round on the palate. When I was looking back at my tasting notes for this one, I wrote, "I want to be alone with this one." No idea what I was thinking, but you can infer that I really liked it.

SHICHIDA JUNMAI DAIGINJO

Brewer: Tenzan Shuzo
Prefecture: Saga
SMV: +2

I wasn't exaggerating when I called Shichida Junmai Daiginjo "the definition of delicate and elegant" in my initial tasting notes on this particular offering. There's some stone fruit up front, specifically a bit of cherry. Its flavor is quite balanced, and its mouthfeel is soft and silky. Did I mention how much I love Yamada Nishiki rice?

AMABUKI STRAWBERRY

Brewer: Amabuki Shuzo
Prefecture: Saga
SMV: 0

No, there aren't any actual strawberries in this nihonshu. But this junmai ginjo does have a faint berry-like character thanks to the use

of a yeast that comes from strawberries. A few whiffs will also reveal
a hint of melon and a touch of citrus. There's also some slight burned
sugar and cooked rice notes that pop up. The mouthfeel has some
moderately creamy elements, so I guess you can say this is like strawberries and cream (it's the only way to be).

DASSAI 39

Brewer: Asahi-Shuzo
Prefecture: Yamaguchi
SMV: +3

Dassai 39—39 being its seimaibuai—is gloriously no-nonsense, as far
as elegant junmai daiginjos go, with a delicate interplay of pineapple,
green apple, and the more floral elements of honey. It has a round,
silky-smooth texture that really makes you want to be friends with 39.

DASSAI 50 SPARKLING NIGORI

Brewer: Asahi-Shuzo
Prefecture: Yamaguchi
SMV: 0

The fine bubbles temper the sweetness a bit, making this nigori seem
a bit drier than it actually is. It has some of the best elements of a
medium-dry Champagne and even a funky Spanish cider, while its dis-
cernible creaminess never lets you forget it's a nigori sake. The finish is
pleasantly tart and dry.

GASSAN JUNMAI

Brewer: Yoshida Shuzo
Prefecture: Shimane
SMV: +1

Shout-out to my friend Christopher Pellegrini, a Vermont-raised
American expat who's lived in Tokyo for most of the twenty-first century, for selecting this one from the Japanese-only menu at a bustling
izakaya in the city's Shinjuku district. (It happened to be Thanksgiving
back in the United States, and instead of turkey and stuffing,
Christopher, my wife, Craige, and I feasted on sashimi and yakitori.)

I got a moderately complex mix of tropical fruit (under-ripe banana, maybe?) and vanilla pudding on the nose of the fairly crisp junmai.

BESSEN JUNMAISHU

Brewer: Hishiya Shuzouten
Prefecture: Miyako
SMV: +3

Sometimes you feel like a nut. If you're experiencing one of those times, this is the nihonshu for you. There's more than a generous helping of roast and toast in this grainy, nutty, umami-dominant junmai. You won't want to drink it too cold; it should be served no more than a few degrees below room temperature. You can also warm it up a little more to accentuate some of that often-elusive umami.

SUIGEI TOKUBETSU JUNMAI

Brewer: Suigei Shuzo
Prefecture: Kochi
SMV: +6.6

Subtle melon, banana, and rice are the aromas that come to mind in this light, clean, and somewhat watery tokubetsu junmai. It's quite drinkable and dynamic enough to pair with a wide range of foods, especially those at the earlier part of the meal.

HOUSHUKU GINJO

Brewer: Nara Toyosawa
Prefecture: Nara
SMV: +3

The dry and crisp ginjo is light on flavor complexity or nuance, but if you twisted my arm, I'd say that a little melon pokes through every once in a while. This is another one of those no-nonsense ginjos if you just want to enjoy a sake and not overanalyze it (I really need those sometimes).

TATEYAMA HONJOZO

Brewer: Tateyama Shuzo
Prefecture: Toyama
SMV: +5

Rich is the term that immediately comes to mind when you nose and sip this honjozo. I get hints of ice cream–filled mocha, banana, custard, and maybe even a little cream puff—but in a dry kind of way.

HANATOMOE ETERNAL SPRING

Brewer: Miyoshino Shuzo
Prefecture: Nara
SMV: +3

There's a good deal of farmhouse funk in a glass of this stuff with some stinky cheese elements, artfully balanced by sour plum, cherry, and apple notes that make it truly something wild and a spring you never want to end.

GASSAN MOUNTAIN MOON

Brewer: Yoshida Shuzo
Prefecture: Shimane
SMV: +4

It's junmai ginjos like this that make me wish *robust* had some better synonyms. On the nose, there's a tremendous amount of tropical fruit—most notably pineapple. There's not too much sharpness on the texture, as it's pretty fluffy and viscous, with a delightfully protracted finish.

HOUSYUKU

Brewer: Toyosawa Honten
Prefecture: Nara
SMV: +5

The aroma is fairly timid, but when it does emerge, there's a bit of creamy vanilla yogurt. As confectionery as that may sound, the finish is very clean and dry.

177

TSUKASABOTAN JUNMAI

Brewer: Tsukasabotan Shuzo
Prefecture: Kouchi
SMV: +7

I'll always remember Tsukasabotan Junmai as a bittersweet experience—not in the literal, flavor sense, but in the emotional one. It will forever be associated with leaving Japan, as it was the last sake I drank at the tail end of my 2017 Japanese sake tour. It was available in Narita Airport's United Club (pour your own!), and I drank it minutes before I boarded the plane that would take me back home. There's an engaging interplay of citrus and sweet, cooked rice—and perhaps a little cream—on the nose, followed by a super dry and clean finish. It's not going to change the world, but it does more than hold its own against the countless sakes I had tasted in the preceding 10 days.

MUSICAL SLIPPERS

U nited Airlines Flight 78 just breached that 10,000-foot threshold at which the flight attendant lets us know over the intercom that it's okay to use large portable electronic devices. It's music to my ears, as I've been eager to wrap up this book and I thought there'd be no more fitting way to do it than on a 12-hour flight home from a sake adventure through Japan.

Well, that's a bit of a lie. I was really hoping to get this part written about 30 hours earlier on the Shinkansen—the world-famous Japanese bullet train that travels at about half the speed of a commercial jet-liner—between Kyoto and Tokyo. But I was at a loss for words then. I was still reeling from my journey through two of the three major epicenters of nihonshu culture, Kyoto's venerable Fushimi District and the Niigata Prefecture (I skipped the other member of the holy trinity, Kobe's Nada district, this time since I had visited it on a previous trip to Japan).

I credit a pair of magic slippers with freeing me from my writer's block. You're probably aware that when you stay at most hotels throughout Japan, you're supposed to take off your shoes when you enter your room. There's usually a pair (or pairs) of slipper sandals waiting for you if you don't want to walk around in socks. You're expected not to steal them. But if you have kleptomaniacal tendencies, you'd be better off resting your head at a lodging property whose rooms include cheap, practically disposable footwear that you're encouraged to take home with you. I typically don't take them up on the offer, but when I spent the last of 10 nights in Japan at the Stay Tokyu, I realized they'd be a godsend on the flight home. I'm never one to take my shoes off on planes—and I silently curse people who do—but keeping Doc Martens on for the duration of the 12-plus-hour journey to Tokyo wasn't the sort of uncomfortable experience I was eager to replicate on the voyage home. The moment I put these thin, ill-fitting quasi-sandals on, I was jolted back to the eight breweries that invited me in to get a rare glimpse at their inner workings. Over the course of three full days of tours, I estimate that I wore around 80 pairs of slippers (and I'm being conservative in that assessment). Not only is it customary to remove your shoes in hotel rooms and certain traditional restaurants and izakayas—especially if there's a tatami room involved—but it's also quite common to do so when attending a business meeting. And if that business happens to be a beverage production facility, expect to change slippers at least a half dozen more times before you leave. Practically every new room you enter—be it the kojimuro, the shubo room, or the large hanger-like area that houses the storage tanks—you'll be swapping one set of footwear for another.

Of course, this isn't a matter of custom, but one more of hygiene. It's easy to contaminate any stage of the sake-making process with microbes that can really undermine so much hard work. At one brewery I was even instructed to remove my watch before we embarked on the tour. "Oh, right," I said, "it could fall in one of the tanks." "No," the CEO replied, "watches can carry a lot of germs."

The odd thing about this is that such fastidiousness seems to be unique to fermented beverage production in Japan. I've been on hundreds of beer brewery, distillery, and winery tours on multiple continents, and there's always a rigorous sanitation regimen. But none ever came anywhere close to the procedures in place at a typical sakagura. The absence of koji in other types of adult beverage production is obviously part of the reason that non-sake, non-shochu-producing facilities are a little more laid back about decontamination.

But an even bigger part of it, I've come to learn, is just the level of respect that nihonshu commands. The youngest of the breweries I visited in Niigata and Kyoto, Musashino Shuzo, was a little more than 100 years old, currently in the hands of the fourth generation of the founding family. "We're the babies," the owners would say. And it's understandable that they'd feel that way when their peers down the road are the tenth- or eleventh-generation descendants of their own companies' early-seventeenth-century founders. There's so much family history and heritage within the brewery walls that you can almost see the weight of it resting on the shoulders of the modern-day stewards. And every time I put on yet another pair of slippers, I felt like I was playing some microscopic role in helping them preserve the legacy of their forebears.

And now that I've cumulatively walked through a couple of millen-nia worth of brewing tradition in a mere handful of days, I suspect that every sip of sake I take from here on out will taste better than it ever has before. Now, if only this plane would hurry up and land so I can get into the cache of bottles I'm bringing back with me (please be gentle, baggage handlers).

And with that, I say to you, a very hearty kanpai!

GLOSSARY

Bone up on your nihonshu-related vocabulary.

»→ Arabashiri (ah-ra-bah-she-ree): In the traditional fune pressing method, the initial sake that flows freely from the moromi/mash bags before actual pressure is applied.

»→ Aru-ten (a-roo-ten): The process of adding distilled alcohol to the sake mash to make honjozo.

»→ Aspergillus awamori: Black koji or kura-koji, typically used to produce the Okinawan spirit awamori, as well as shochu.

»→ Aspergillus kawachi: Whit koji or shiro-koji, typically used to produce the distilled spirit shochu.

»→ Aspergillus oryzae: Yellow koji or ki-koji, the type used in producing sake.

»→ Awamori (Ah-wah-mor-ee): A shochu-like beverage distilled from Indica rice native to Okinawa.

»→ Binzume yosui (bin-zoo-meh-yo-soo-ee): Water used for post-production activities like bottle and production equipment cleaning as well as diluting the sake (warimizu).

»→ Choko (Cho-ko): A small sake cup, traditionally made of earthenware but also available in porcelain or lacquer.

»→ Choshi (Cho-she): A tea kettle–like container used for pouring sake into cups.

»→ Daiginjo (Die-ghin-joe): Sake that uses rice with at least 50 percent of its outer layers polished away.

»→ Dakidaru (dah-kee-dar-oo): A container, often made of wood, that's filled with hot water and immersed in the yeast starter (shubo)

183

during the yamahai process to help raise the temperature to one ideal for lactobacillus to produce lactic acid.

»→ Dekoji (Deh-ko-jee): The seventh and final step in the rice koji production process.

»→ Dewasansan (Deh-wah-sahn-sahn): A variety of sake rice grown primarily in Yamagata Prefecture.

»→ Doburoku (do-boo-ro-koo): Completely unfiltered and milky white, it's one of the earliest forms of sake. It's still produced, to a small extent, today.

»→ Echigo Toji (Eh-chee-go-toe-jee): Based in Niigata Prefecture (formerly known as Echigo), it is of the most respected guilds for sake masters.

»→ Fukurozuri or fukuro-tsuri (foo-koo-ro-zoo-ri): A method of sake pressing during which the liquid slowly drips through hanging cotton bags, separating from the solids.

184

»→ Fukuryusui (foo-koor-ee-oo-soo-ee): A prized water source derived from Mount Fuji snowfall, it's known to produce a sake that is quite crisp and has a soft texture.

»→ Fune (foo-nay): A large wooden box used for pressing sake. The fermented mash is poured into cotton bags, which sit at the bottom of the fune while its lid is pushed down, forcing the liquid out of the bags and through a hole in the box.

»→ Futsu-shu (foot-sue-shoo): Common, everyday, below-premium sake.

»→ Gaikonainan (guy-ko-nigh-nahn): The desired quality of the rice when steamed, it means firm on the outside and soft on the inside (kind of like "al dente" in pasta).

»→ Genshu (ghen-shoo): Undiluted sake; it foregoes warimizu and often can be as high as 20 percent ABV (typical sake is diluted down to about 15 percent ABV).

»→ Ginjo (Ghin-joe): Sake that uses rice with at least 60 percent of its outer layers polished away.

➻ Ginpu (Ghin-poo): A premium sake rice variety that grows primarily in Hokkaido.

➻ Gohyaku Mangoku (go-ya-koo-mahn-go-koo): A variety of premium sake rice developed in 1938; it tends to produce light sake with little complexity. It grows in Niigata as well as in Fukui, Ishikawa, and Toyama Prefectures.

➻ Gokosui (Go-ko-soo-ee): A famous water source from the Fushimi region of the Kyoto Prefecture, it's known to yield sweeter sake.

➻ Guinomi (Gwee-no-me): A midsize sake cup, larger than a choko and typically used for chilled sake.

➻ Hatsuzoe (hot-soo-zo-eh): The first day of sandan jikomi (three-step brewing process) when about 20 percent of the total steamed rice, rice, and rice koji are added to the shubo.

➻ Hattan Nishiki or Hattannishiki (ha-tahn-nee-shee-kee): A variety of premium sake rice from the Hiroshima Prefecture, it's among the easier strains to cultivate as it doesn't grow very high and therefore is more resistant to wind. Hattan Nishiki–based sakes tend to be light and refreshing.

➻ Hi-ire (Hee-eeray): The pasteurization process.

➻ Hikikomi (He-Kee-ko-mee): The first step in the rice koji production process.

➻ Honjozo (hon-joe-zo): Sake with distilled alcohol added to the mash.

➻ Isshobin (ee-sho-bin): A large, 1.8-liter bottle that holds the equivalent of 10 servings of sake.

➻ Izakaya (ee-zah-keye-ah): Literally "stay sake shop," a casual Japanese pub known for serving a variety of small plates (think Japanese tapas) to consume along with one's adult beverages; red lanterns often hang outside traditional izakayas.

➻ Janomeno Kikichoko (Jah-no-meh-no Kee-Kee-cho-ko): A cup that sake brewers use for analysis with blue and white circles at the bottom that enable brewers to assess the color.

»→ **Jizake (Jee-zah-keh):** Regional, artisanal sake, distinguished from the large national brands (think the Japanese sake version of craft beer).

»→ **Joso (Joe-so):** The process of pressing sake to remove the solids from the liquid.

»→ **Juku-shu (Joo-koo-shoo):** One of four flavor and aroma classifications created by the Sake Service Institute. Aged sakes that are intense in aroma and more subtle in flavor fit into this category.

»→ **Junmai (Joon-my):** Pure rice sake containing no added distilled alcohol.

»→ **Junmai ginjo (Joon-my-geen-joe):** Pure rice sake containing no added distilled alcohol where at least 40 percent of the rice's outer layer has been polished away.

»→ **Junmai daiginjo (Joon-my-die-geen-joe):** Pure rice sake containing no added distilled alcohol where at least 50 percent of the rice's outer layer has been polished away.

»→ **Jun-shu (joon-shu):** One of four flavor and aroma classifications created by the Sake Service Institute. Jun-shu sakes are known for their strong, rich flavors and aromas.

»→ **Kake Mai (ka-kee-my):** The rice that will be used for the moromi (mash).

»→ **Kan (kahn):** When sake is heated.

»→ **Kanpai/Kampai (Kahn-pie):** Literally "dry cup," it's the Japanese toast equivalent of "Cheers!"

»→ **Kasu (kah-soo):** The solids (lees) separated from the liquid when sake is pressed.

»→ **Katakuchi (kah-tah-hah-chee):** A sake decanter with a wide, open mouth.

»→ **Ki-Ippon (Key-ee-pohn):** Think of it as the sake equivalent of a single malt whisky: a nihonshu brewed and bottled at one location with no addition of alcohol.

»→ Kimoto jikomi (Kee-mo-tow-jee-ko-mee): The older of the two kimoto-kei-shubo methods in which the brewers use the yamaoroshi process, where they're constantly jamming oar-like poles into the mixture to puree the rice and aid saccharification.

»→ Kimoto-kei-shubo (kee-mo-tow-kay-shoo-bo): The traditional method of making the shubo, during which lactic acid–producing bacteria are allowed to propagate on their own—the two methods are: kimoto jikomi and yamahai jikomi.

»→ Kirikaeshi (Keer-ee-keye-esh-ee): The third step in the rice koji production process.

»→ Koji Kin or simply "Koji" (ko-jee-kin): A mold that's inoculated onto rice that produces the enzymes necessary to convert starch to fermentable sugar.

»→ Kojimai (ko-jee-mai): The rice that will be used to cultivate koji.

»→ Koji Muro (ko-jee-more-oh): The room where rice koji is produced; it's kept at a constant, humid temperature of about 95 degrees Fahrenheit (35 degrees Celsius).

»→ Komekouji (ko-meh-ko-gee): Steamed rice after koji has been cultivated on it.

»→ Kongo (cone-go) rolls: Large, spinning grindstones inside the seimaiki that polish rice.

»→ Koshiki (ko-she-kee): A large vat used for steaming rice used in sake making, traditionally made of cedar but now more commonly made of stainless steel.

»→ Koshi Tanrei (ko-she-tan-ray): A variety of shuzo kotekimai that grows in the Niigata Prefecture and used mainly by breweries located there.

»→ Kun-shu (koon-shoo): One of four flavor and aroma classifications created by the Sake Service Institute, Kun-shu sakes tend to be highly fragrant, with lighter, more delicate flavors. Most daiginjo-grade sakes and many ginjo-grade sakes fall under this heading.

»→ Kurabito (Core-ah-bee-toe): A sake brewery worker who works under the toji.

»→ Kyubetsu seido (kyoo-bet-soo-say-do): A system developed in 1943 to separate sake types by class for the purpose of taxation. After periodic revisions, Japan permanently abolished the system in 1992 and replaced it with the practice of taxing sake based on alcohol content.

»→ Lactic acid: Produced by lactobacillus bacteria, it's an acid necessary to allow yeast to propagate when making sake. Yeast can live in such an acidic environment, while unwanted microbes do not.

»→ Masu (Mah-soo): A box, traditionally made of wood but also commonly made of lacquer, that holds 180 milliliters of sake based on the measurement for a single serving of rice. When a server pours sake into it, there's a custom of letting it overflow into a small plate below.

»→ Mizokiri: The stage in rice cultivation when the farmers dig shallow trenches in the rice paddy for proper water distribution.

»→ Miyama Nishiki (Mee-ya-ma-nee-shee-kee): A variety of premium sake rice known to hold up fairly well in the cold in prefectures like Nagano and Akita. The sake it produces tends to be crisp, just like the climate in which the rice grows.

»→ Miyamizu (Mee-ya-mee-zoo): A famous and prized water source from the Nada region of Kobe in the Hyogo Prefecture known for its high minerality, particularly desirable ones (for sake) like phosphorus and potassium.

»→ Mori (More-ee): The fourth step in the rice koji production process.

»→ Moromi (mor-o-mee): The rice-based mash that's fermented to become sake.

»→ Multiple Parallel Fermentation: The process, unique to sake among fermented beverages, by which saccharification and fermentation occur at the same time.

»→ Muroka (moo-roe-ka): Sake that has not been filtered; not to be confused with nigori, as nigori has been filtered. Also, muroka sake can be clear.

»→ **Mushi (moo-she):** This is the steaming of rice that will eventually become sake. Where other fermented beverages like beer are mixed with water, sake gets its moisture from steaming.

»→ **Nakaboshi (na-kah-bo-she):** The stage in rice cultivation, following mizokiri, when water is drained.

»→ **Nakashigoto (Na-kah-she-go-toe):** The fifth step in the rice koji production process.

»→ **Nakazoe (nah-kah-zo-eh):** The third day of sandan jikomi (three-step brewing process) when about 30 percent of the steamed rice, rice koji, and water are added to the moromi.

»→ **Nama:** The term for "raw," which usually serves as a prefix for various types of sake that forego one or both of the rounds of pasteurization typical for sake. *Nama* is often used as a catch-all term for unpasteurized sake, regardless of which hi-ire steps it bypasses.

»→ **Namachozo (nah-mah-cho-zo):** Sake that bypasses the first round of pasteurization, post-filtration, and pre-tank-storage, but goes through the round just prior to bottling.

189

»→ **Namazake (nah-mah-zah-kay):** Sake that's completely unpasteurized, sometimes referred to as "nama-nama."

»→ **Namazume (nah-mah-zoo-meh):** Sake that goes through the first pasteurization before tank storage but bypasses the stage prior to bottling.

»→ **Nigori or nigorizake (nee-gore-ee or Nee-gore-ee-zah-keh):** A white, cloudy style of sake that has gone through a coarse filtration, leaving many of the solid rice particles in the liquid.

»→ **Nihonshu (Nee-hon-shoo):** Literally "Japanese sake," a term that's used interchangeably with "sake."

»→ **Omachi (oh-mah-chee):** A variety of premium sake rice grown in the Okayama Prefecture that's known to produce sake with some earthy qualities.

»→ Odori (oh-door-ee): The second day of sandan jikomi (three-step brewing process) when no additional rice, rice koji, or water are added to allow the yeast to propagate.

»→ Ori (or-ee): Sediment that settles to the bottom of the tank in which the fresh-pressed sake sits for about 10 days before it's filtered.

»→ Oribiki (or-ee-bee-kee): The process of removing the ori.

»→ Roka (ro-kah): Filtration.

»→ Sakamai (sah-kah-my): Another term for rice used to make sake (*see also* shuzo kotekimai).

»→ Sake Meter Value (SMV): A sweetness/dryness scale for sake. Negative numbers signify sweetness; positive numbers represent dryness.

»→ Sandan jikomi (san-dan-jee-ko-mee): The three-step-brewing process, during which rice, rice koji, and water are added at three different intervals over the course of four days to ensure an optimally acidic environment for yeast to thrive and other, unwanted microorganisms to be fought off. Adding it all at once would dilute the acidic environment.

»→ Seimaibuai (say-my-bwhy): The polish ratio of rice used in sake making, expressed in a percentage, representing the portion of the grain that remains. For example, a seimaibuai of 60 percent means 40 percent of the grain has been milled away.

»→ Seimaiki (say-my-kee): The vertical rice polishing machine.

»→ Seishu (say-shoo): The official, legal name for sake in Japan. It means "clear sake."

»→ Senmai (sen-my): The stage in the production process, post-polish, when the rice is washed to remove any lingering particles.

»→ Senryuju (sen-ree-oo-joo): The weight of 1,000 grains of rice. Typically, sake rice has a senryuju of 25 to 30 grams, a weight that's slightly higher than that of table rice.

»→ Shimaishigoto (shee-my-ee-shee-go-toe): The sixth step in the rice koji production process.

➤ Shinpaku (shin-pah-koo): The dense, white, starchy center of the rice kernel.

➤ Shinseki (shin-seh-kee): The stage in the production process when the rice soaks to absorb enough water to prepare it for steaming.

➤ Shochu (sho-choo): A clear, distilled beverage native to Japan that, like sake, uses koji in its production process.

➤ Shubo (shoo-bo): The yeast starter, sometimes called moto, consisting of steamed rice, rice koji, and water to which yeast is added.

➤ Shuzo (shoo-zo): A company that produces alcohol; a brewery (if sake or beer) or a distillery (if shochu).

➤ Shuzo Kotekimai (shoo-zo-ko-teh-kee-my): Rice designated for making sake. There are more than 100 different varieties.

➤ Shuzo-yosui (shoo-zo yo-soo-ee): Water used in the production of sake.

➤ Sokujo-kei-shubo (so-coo-jo-kay-shoo-bo): A modern process of developing the shubo in which commercially available lactic acid is added, cutting the shubo time in half.

➤ So-shu (So-shoo): One of four flavor and aroma classifications developed by the Sake Service Institute. Primary characteristics of so-shu sake is light flavor and subdued aroma. So-shu, generally, is considered refreshing.

➤ Souhazegata (soo-ha-zeh-gah-ta): A type of rice koji that covers the entire surface of the rice kernel and penetrates deep into the center of the grain. It's used primarily if a full-bodied sake is desired.

➤ Sugidama (soo-ghee-dah-ma): A sphere made from small tree branches that have been bound together. Brewers traditionally hung the green shrubbery ball at the beginning of the brewing season.

➤ Taru zake: Sake that's aged in wooden—most often cedar—barrels. Like whisky and other barrel-aged beverages, taru picks up some of the flavor and color of the wood.

》→ Tobin (Toe-bin): An 18-liter bottle into which the sake is collected during the fukorozori pressing method.

》→ Tobingakoi (toe-bin-gah-koi): The process of capturing the drops of sake from fukurozuri into a tobin.
》→ Toji (toe-jee): The sake brew master.

》→ Tokkuri (toe-koo-ree): A carafe used for serving sake.

》→ Tokomomi (Toe-ko-mo-mee): The second stage of rice koji production.

》→ Tokubetsu honjozo (toe-koo-bet-soo-hon-joe-zo): A special form of honjozo often with a seimaibuai of 60 percent versus the typical 70 percent for premium honjozo.

》→ Tokubetsu junmai (toe-koo-bet-soo-joon-my): A pure rice sake made through a special brewing process.

》→ Tomezoe (toe-meh-zo-eh): The fourth day of sandan jikomi (three-step brewing process) when the remaining 44 percent (give or take) of the steamed rice, rice koji, and water are added to the moromi.

》→ Tsukihazegata (soo-kee-ha-zeh-gah-ta): A type of rice koji that penetrates to the center of the kernel but only sparsely appears on the surface; used for sakes that are a bit smoother and lighter in flavor.

》→ Warimizu (wah-ree-mee-zoo): The production stage in which sake is diluted with water.

》→ Yabuta (ya-boo-ta): The most popular brand of automatic sake pressing machine.

》→ Yamada Nishiki or Yamadinishiki (ya-ma-da-nee-shee-kee): A famous, high-quality type of shuzo-kotekimai.

》→ Yamahai jikomi (ya-ma-high-jee-ko-mee): One of the two kimoto-kei-shubo methods, it was developed when scientists at the National Institute of Brewing Research determined that jabbing the rice with a stick—the yamaoroshi process—was unnecessary for saccharification.

》→ Yamaoroshi (Ya-more-oh-shee): The labor-intensive process of aggressively poking the shubo with a long, oar-like rod, once

192

believed to be necessary to facilitate saccharification of starch in the rice.

»→ Yongobin (yon-go-bin): A 720-milliliter sake bottle.

»→ Yontotaru (Yon-toe-tar-oo): A traditional 72-liter sake barrel that's mostly decorative these days but is also used during celebrations.

»→ Yukimuro (yoo-key-more-o): A room that holds snow, used to maintain a low temperature when aging sake.

ACKNOWLEDGMENTS

Writing this book has been one of the greatest adventures of my life—and that's saying a lot considering my first book was called *The Year of Drinking Adventurously*. And it was by no means a solo adventure, as there's an incredibly long list of people, organizations, breweries, importers, bars, and restaurants that made this project happen.

First I must thank Todd Bottorf, Jon O'Neal, Stephanie Bowman, and everyone else at Turner Publishing. This is the third book of mine that Turner has published, and I'm constantly amazed that they continue to do so.

Now, on to those on the sake-making/drinking/selling/importing/distributing side.

Huge thanks to Timothy Sullivan of Urban Sake for agreeing to be an extra set of eyes on my manuscript.

I also owe a big shout out to Jamie Graves, someone who was not only instrumental in connecting me with the right people in Japan, but who has also been my go-to sake guru since my first book. I hope to be as knowledgeable about sake (not to mention the Japanese language) as Jamie one day.

Jamie put me in touch with Ataru Kobayashi, who became my incredibly generous guide through the Niigata sake scene. He organized an extensive itinerary throughout the prefecture, involving seven breweries, countless tastings, and several meals that were simply *oishi*. It was the best hands-on nihonshu education I've ever had.

Speaking of Niigata, I must express my deepest gratitude to the following: Eriko Mukoda and everyone at Kirinzan Shuzo, Takako and Tomoyuki Shigeno and their team at Kinshihai Shuzo, Makoto and Yutaka Furusawa and everyone at Matsunoi Shuzojo, Hisashi and Hajime Kobayashi and all of the folks at Musashino Shuzo, Shunji Odaira at Midorikawa Shuzo, and everyone at Aoki Shuzo and Kiminoi Shuzo. You all gave Craige and me such a warm welcome during our time in your snowy, picturesque prefecture, and this book is so much better for it.

And I can't talk about Niigata without raising a choko to Ponshukan at Niigata City's main train station. The coin-op (well, technically, token-op) tasting wall is worthy of its own book!

Moving on to Kyoto, a huge "kanpai!" to Yuji Fukai and Tamanohikari Shuzo for showing us around on a Saturday morning, and to Aburacho for continuing to offer one of the best crash courses in the prefecture's brewing scene. A thanks is also owed to Torisei for not only having a menu full of tasty Fushimi nihonshu, but some incredibly delicious yakitori.

A few other folks in Japan who have been instrumental in this and other writing projects: Stephen Lyman, Christopher Pellegrini, and Noriyuki (Nori) Yamashita. By the time you guys read this, I promise I will have taken the Shochu Adviser exam. I meant to sooner, but had to put it on hold to get sake certification. To everyone else reading this, if you ever find yourself in Kumamoto, make sure you visit Nori's shochu bar, Glocal Bar Imo Vibes.

Back in the United States, I want to thank a few people I've subjected to my ongoing research. Whether they knew it or not, when they agreed to meet me at sake bars—instead of more crowd-pleasing venues like alehouses, wine bars, and cocktail lounges—they became co-conspirators in the creation of this tome: Sarah and Giancarlo Annese, Mary Kate and Ben Mack, Kristina and Mike Mansbridge (on both sides of the Atlantic!), Clare and Adam Sivits, Renee Hickerson, Erika Bolden and Beth Gerbe.

This wouldn't be an acknowledgment section of one of my books if I didn't lift my glass to John Holl. He's thrown work my way from every publication he's worked for over the past handful of years. But the best, by far, was the sake instructional video he invited me to star in for Craft Beer & Brewing. Now I just need to get him (and his bowtie) to drink more of this stuff.

A great big thanks to Steve Vuylsteke at SakeOne in Forest Grove, Oregon, for always being kind enough to let me stop by and pick his brain; Dan Ford of Blue Current Brewery in Kittery, Maine, for meeting me on a busy, late Friday afternoon to show me around his digs, and to Marc Hughes at Gaijin 24886 in Denver. I must admit, I've been getting a bit of GABF fatigue over the past few years, but I'm thankful that this operation has given me an additional reason to head to the Mile High City each fall. I'd also like to extend a similar expression of gratitude toward Byron Stithem of Proper Sake in Nashville. He's doing God's work there in Music City (also, a thanks for hosting the event for my previous project, *The Drinkable Globe*, when I was in town).

And since we're talking about American sake makers, I want to direct the spotlight toward Dovetail in Boston, Texas Sake Company in Austin, Jake Myrick of Sequoia Sake in San Francisco, Setting Sun and Kuracali in Southern California, and Brooklyn Kura. By the time you read this, I hope that list will have gotten much longer. There are great things happening in the United States, and I'm excited to see how it continues to develop.

Beyond the breweries, I want to acknowledge the bars, restaurants, and izakayas that are educating drinkers every day: Decibel, and Sakagura in New York; Bamboo Sushi and Zilla in Portland (special thanks to Kate Koo!); Izakaya Seki, Daikaya, Sushi Capitol, Umaya, and Sushi Taro in Washington, DC; Izakaya Den in Denver; and Murasaki (especially Ronnie Prince) and Booze Box in Chicago. Also, double thanks to Marcus Pakiser for the wonders he's worked in growing the market for sake in the Pacific Northwest. Speaking of growing, it cannot be understated how much of a game-changer the grain Chris Isbell is growing in Arkansas will be for the American sake industry.

Finally, without the Sake School of American and Mutual Trading and my teachers, Toshio Ueno, Eric Imamura, and Sara Guterbock, I would have neither the knowledge nor the confidence to write *Sakepedia*.. I started 2017 with this crazy idea to get some sort of formal training related to the beverage I'd been most passionate about for about a decade. I thought I'd be done after passing the Certified Sake Adviser test in February, but then I figured, What the hell, I might as well go for the next level, Certified Sake Sommelier (aka International Kikisake-shi). The prospect of taking it terrified me. The Adviser class was more of the book-learning sort, but the Sake Sommelier exam had a significant practical component. I'll admit that I didn't initially have the confidence in my own taste buds to think I'd come close to passing it. In fact, about an hour into the three-day intensive program, I texted Craige to tell her, "I think I made a mistake, I'm way out of my depth."

But finishing the course and passing the exam made me realize how subjective an experience sake is. It speaks to all of us in different ways (the term "Kikisake-shi" actually tells us to "listen to the sake") and we should all form our own relationship with the beverage. And when I say "all of us," I mean it. Sake is one of the most welcoming, inclusive beverages there is—it's for everyone!

Among that "everyone" is my favorite "one," my constant kanpai companion, Craige. I was so glad she able to take a more hands-on role in the research for this book and that she continues to support my crazy, creative flights of fancy, regardless of how much it keeps me away from home. I'm also glad we finally have a beverage that we've fallen in love with together. It's great that we can finally agree on what to drink—though I'm convinced that one day she'll start liking whisky. Dare to dream.

197

RESOURCES

»→ Interviews

Auffrey, Richard. Phone interview, 13 Dec 2017.
Doughan, Brandon. In-person interview, 2 March 2018.
Ford, Dan. In-person interview, 8 Dec 2017.
Fukai, Yuji. In-person interview 2 Dec 2017.
Furusawa, Makoto. In-person interview, 28 Nov 2017.
Graves, Jamie. Phone interview, 8 Jan 2018.
Hembree, Josh. Phone interview, 18 December 2017.
Hughes, Marc. Phone interview, 6 Jan 2018.
Isbell, Chris. Phone interview, 7 Feb 2018.
Kobayashi, Ataru. In-person interviews, 27-29 Nov 2017.
Kobayashi, Hisashi. In-person interview, 29 Nov 2017.
Koo, Kate. Phone interview, 7 Feb 2018.
Myrick, Jake. Phone interview, 12 Jan 2018.
Odaira, Shunji. In-person interview, 28 Nov 2017.
Pakiser, Marcus. Phone interview, 9 Feb 2018.
Polen, Brian. In-person interview, 2 March 2018.
Prince, Ronnie. Phone interview, 23 Jan 2018.
Shigeno, Takako. In-person interview, 27 Nov 2017.
Stithem, Byron. In-person interview, 26 Jan 2018.
Vuylsteke, Steve. In-person interview, 24 Jan 2018.
Wight, Trevor. Phone interview. 12 Dec 2017.

»→ Sake Brewery Tours

Aoki Shuzo, Minami-Uonuma, Japan, 28 Nov 2017
Blue Current Brewery, Kittery, Maine, USA, 8 Dec 2017
Brooklyn Kura, Brooklyn, N.Y., USA, 2 March 2018
Kinshihai Shuzo, Gosen, Japan, 27 Nov 2017.
Kirinzan Shuzo, Tsugawa, Japan, 27 Nov 2017.
Kiminoi Shuzo, Myoko, Japan, 29 Nov 2017.
Matsunoi Shuzojo, Tokamachi, Japan, 28 Nov 2017.
Midorikawa Shuzo, Uonuma, Japan, 28 Nov 2017.
Musashino Shuzo, Joetsu, Japan, 29 Nov 2017.
Proper Sake, Nashville, Tennessee, USA, 26 Jan 2018.
SakéOne, Forest Grove, Oregon, USA, 24 Jan 2018.
Setting Sun Sake, San Diego, California, USA 11 Aug 2018
Tamanohikari Shuzo, Kyoto, Japan, 2 Dec 2017.

⇥ Courses
Sake Adviser Certification, Sake School of America, Secaucus, New Jersey, February 2017.
Sake Sommelier Certification, Sake School of America, Los Angeles, California, August 2017.

⇥ Online resources
John Gauntner's Sake World, Sake-World.com
Niigata Sake Selections, NiigataSake.com
PassionateFoodie.blogspot.com, Richard Auffrey, editor
SakeSocial.com
UrbanSake.com, Tim Sullivan, Editor
Hidaka, Masahiro and Takeo, Yuko. "Fukushima Looks to Top-Tier Sake to Beat Stigma, Lift Economy." *Bloomberg*, 28 Dec 2017, https://www.bloomberg.com/news/articles/2017-12-28/fukushima-looks-to-top-tier-sake-to-beat-stigma-lift-economy.
Fukushima-Sake.com
Sake-Hiroshima.com
True-Sake.com
VisitGunma.jp

⇥ Festivals & Trade Events
Sake Summit, San Francisco, California, 9 Sept 2017
Joy of Sake, New York, New York, 15 June 2018.
PDX Sake Festival, Portland, Oregon, 27 June 2018.

ABOUT THE AUTHOR

Jeff Cioletti's tenure in liquid literacy has exposed him to some of the best libations the world has to offer and given him access to the producers and purveyors of such fine refreshments. He combines his love of drink with a passion for travel and one usually involves the other.

He served for fourteen years as an editor at Beverage World magazine, including eight years as editor in chief. He's also the author of the books, "The Drinkable Globe," "The Year of Drinking Adventurously" and "Beer FAQ."

Jeff is the founder of DrinkableGlobe.com and host of The Drinkable Globe Podcast. He's a frequent contributor to publications including Artisan Spirit Magazine, Beverage Media, BevNet, Beverage Industry, The Takeout, SevenFifty Daily and CraftBeer.com. Additionally, he's certified as an International Kikisake-shi (sake sommelier) by Sake Service Institute International and the winner of four North American Guild of Beer Writers awards.

CPSIA information can be obtained
at www.ICGtesting.com
Printed in the USA
BVHW031450100319
542257BV00004B/11/P